QK 99 .A1 A2 1985
Abbe, Elfriede Martha, 1919-
  The fern herbal : including the ferns,
the horsetails, and the club mosses

DATE	ISSUED TO

QK 99 .A1 A2 1985
Abbe, Elfriede Martha, 1919-
  The fern herbal : including the ferns,
the horsetails, and the club mosses

The Fern Herbal

THE FERN HERBAL

*including the Ferns, the Horsetails
and the Club Mosses*

Written and illustrated

by

ELFRIEDE ABBE

Comstock Publishing Associates
A division of Cornell University Press
Ithaca and London

Copyright © 1981 by Elfriede Abbe

All rights reserved. Except for brief quotations in a review, this book, or parts thereof, must not be reproduced in any form without permission in writing from the publisher. For information address Cornell University Press, 124 Roberts Place, Ithaca, New York 14850.

Reprinted, corrected edition first published 1985 by Cornell University Press.

Published in the United Kingdom by Cornell University Press Ltd., London.

Limited edition of 150 copies published in 1981 by the Press of Elfriede Abbe.

International Standard Book Number 0-8014-1718-X
Library of Congress Catalog Card Number 84-45439
Printed in the United States of America
Librarians: Library of Congress cataloging information appears on the last page of the book.

The paper in this book is acid-free, and meets the guidelines for permanence and durability of the Committee on Production Guidelines for Book Longevity of the Council on Library Resources.

This book is reprinted by offset from the limited edition printed and published by the Press of Elfriede Abbe. The type faces are Goudy Deepdene Italic and Goudy Oldstyle and Italic. The illustrations in the original edition were wood engravings printed from the original blocks, and drawings were printed from metal plates. All of the type was hand set, and the sheets were printed on a 10 x 15 Chandler and Price platen press, manually operated.

Contents

Introduction
The Fern Plant 1

I. THE FERNS
   * Commonly called fern, but does not belong to the Fern Order.

Adiantum capillus-veneris L.	Southern maidenhair	7
Adiantum pedatum L.	Maidenhair	13
Asplenium adiantum-nigrum L.	Black maidenhair	17
Asplenium ruta-muraria L.	Wall rue	21
Asplenium trichomanes L.	English maidenhair	25
Athyrium filix-femina (L.) Roth	Lady fern	29
Botrychium lunaria (L.) Sw.	Moonwort	33
Botrychium virginianum (L.) Sw.	Rattlesnake fern	37
Ceterach officinarum Lam. & DC.	Ceterach	41
* *Comptonia peregrina* (L.) Coulter	Sweet-fern	42
Cibotium barometz (L.) J. Smith	Vegetable lamb	43
Dennstaedtia punctilobula (Michx.) Moore	Hay-scented fern	47
Dryopteris filix-mas (L.) Schott	Male fern	51
Dryopteris fragrans (L.) Schott	Fragrant fern	55
Dryopteris spinulosa (O. F. Muell.) Watt	Spinulose wood fern	59
Matteuccia struthiopteris (L.) Todaro	Ostrich fern	60
Osmunda claytoniana L.	Interrupted fern	61
Osmunda regalis L.	Royal fern	65
Ophioglossum vulgatum L.	Adder's tongue	69
Phyllitis scolopendrium (L.) Newm.	Hart's-tongue fern	73
Polypodium vulgare L.	Polypody	77
Pteridium aquilinum (L.) Kuhn	Bracken	81

II. THE HORSETAILS — 87

Equisetum arvense L.	Field horsetail	91
Equisetum fluviatile L.	Great horsetail	92
Equisetum hyemale L.	Scouring rush	93
Equisetum sylvaticum L.	Wood horsetail	94

III. THE CLUB MOSSES — 97

Lycopodium clavatum L.	Common club moss	99
Lycopodium complanatum L.	American ground pine	101

APPENDIX

* *Asparagus plumosus* Baker	Asparagus fern	103
* *Myrrhis odorata* Scop.	Sweet fern	103
Pellaea mucronata D. C. Eaton	Bird's-foot fern	103

Introduction

🌿 *Among the ferns and their relatives in their quality as herbs* history reveals a great diversity of uses as well as abrupt changes in the status and importance of the different species. Of the ferns with which humanity has been concerned for at least two thousand years, the one formerly of most versatile use is now considered mainly as a weed. Another with reputation of equal antiquity among herbalists and physicians has retained, in spite of the modern drug industry, a unique place as a specific remedy. Three others, say some modern authorities, ought not to be classified as ferns at all, but should form a separate order in the plant kingdom. Into the contemporary world of the 'gourmet' have come several other ferns, their primitive or exotic use as food a twentieth-century rediscovery. Two as fragrant herbs have been neglected and should be better known. One of the ferns' relatives has peculiar spores which are used in fireworks. Visually ferns hold their exclusive place among the most beautiful of herbs, including the rose, the chrysanthemum, the lily and the iris, by the elegant forms of their leaves, which often act as foils for the color of a flower. As Thoreau said, 'Nature made ferns for pure leaves, to see what she could do in that line.'

Descendants of the first ferns are found today in warm, moist mountain regions of South and Central America. Some species reach great height and are called tree ferns. They have been used by primitive peoples since prehistoric ages for food and building materials.

In the temperate regions of Europe, America and parts of Asia grow the herbaceous ferns and their kindred, the horsetails and club mosses, small in size compared to their ancient ancestors or modern tropical relatives. These are the ferns of history, familiar to mankind as herbs since ancient times. Inquiring and inventive Greeks, practical Romans, the troubled seekers of the Middle Ages, the self-assured humanists of the Renaissance, Indians of the

Americas, colonial housewives, all invested the ferns with mystery, focused on them in study and controversy, and put them to practical uses. Through the centuries it is to the cottage that the fern has been closest, from medicine for the sick to the thatch of the roof.

That ferns 'bear neither flower nor seed' has always been commonly observed, and the old herbalists, in perplexity over fern reproduction, repeated the statement endlessly. It was said that ferns originally bore flowers, but on the night of the Nativity, when all other plants bloomed, the ferns did not, and so were forever after doomed to be flowerless. Yet less than fifty years before the Nativity, the fern was praised by the greatest of Roman poets as the best litter for sheep and cattle. Serving its own humble purpose, fern lay on the ground of many a stable on the first Christmas Eve.

Hieronymus Bock, a German botanist born in 1498, misled neither by superstition nor by allegiance to the ancients, was the first to collect spores of bracken, thereby dispelling some of the mystery of 'fern seed.' Culpeper and Parkinson followed his example. The spores were recognized as reproductive elements, but their nature and function were not discovered or understood until much later. Even in the nineteenth century some botanical scholars insisted that spores were seeds and declared any other concept outrageous.

Bock also established the term 'simple' for an individual herb used as the ingredient of a compound medicine, 'die Einfache Erd Gewaechs, Simplicia,' meaning 'the simple growing things of earth.' The word became a verb, as in William Cole's work of 1656, *The Art of Simpling*. Says Oliver Goldsmith:

> Botanists, all cold to smiles and dimpling,
> Forsake the fair, and patiently—go simpling.

Simplicia in Platearius' *Circa Instans* denote not only plants but a variety of medicinal ingredients, animal and mineral, as well as vegetable. Other substances were also included in *simplicia* in Brunschwig's *Liber de arte distillandi*, but the predominance of botanical material accounts for Bock's emphasis on plants, which constituted the principal source of drugs until modern times.

In the following descriptions of herbs several early works are quoted. Two divergent attitudes emerge from the pages of the old botanical writers. One is a firm adherence to mystic doctrines, such as alchemy and astrology. The other is a pragmatic and rational approach, the beginning of scientific method in research. Although these points of view sometimes overlapped, in general each scholar followed a distinct path. As an exponent of quackery, easily the most notorious was Nicholas Culpeper (1616—1654), a self-appointed astrologer and physician, always at odds with the established medical profession, for he pirated the *Pharmacopoeia* of the College of Physicians, adding information of his own and publishing the result as *A Physicall Directory* (1649) and in 1653 as *The English Physition enlarged*. Culpeper proclaimed that all nature is governed by the planets, and he prescribed his botanical remedies accordingly. One of Culpeper's most vigorous attackers was William Cole, a proponent of the Doctrine of Signatures, which held that whatever a plant resembled it was meant to cure, the liver-shaped leaf of hepatica for liver ailments, for example. Cole argued against Culpeper's tenet by reminding the public that plants were created on the third day but planets not until the fourth day. Culpeper's book retained its popularity, however, and appeared in numerous editions.

In the social and intellectual confusion of the Dark Ages after the fall of Rome, medical practitioners clung to the writings of two Greeks, one a philosopher, born in 372 B.C.; the other was a doctor of the first century A.D., who was called Pedanius Dioscorides. He was born in Cilicia, a region of the coast of Asia Minor. Little is known of him, except as author of a work on medicinal substances, mostly plants. Despite inadequate descriptions and queer pictures, the manuscripts of this *De materia medica* were held in almost holy veneration through the Middle Ages and served as the prototype and foundation of later pharmacopoeias.

No little reverence for ancient manuscripts arose from their reputation as magic books of spells. The medicos of ancient Greece and Rome used incantations to accompany treatment, as did their red-skinned contemporaries on an unknown continent in the West when gathering and using healing herbs. The physicians repeated names of plants in all known languages as magic formulas,

the sound perhaps having a soothing effect on the patient. Lists of plant names in Arabic, Greek, Hebrew, Persian and other languages fill numerous pages of the *Herbarium* of Apuleius, a fourth-century A.D. derivative of Dioscorides' work. Monks who copied this and other classical works changed or added prayers to the incantations, but the peasantry continued to shout, 'Vervain, basil, Johnswort, dill!' at witches to counteract evil spells.

So esoteric and seqestered in monasteries were early botanical works, that much knowledge of plants among the laity was passed from one generation to another by word of mouth. Although Dioscorides' manuscript was translated by John Goodyer and shared with his friend, John Gerard, the result was something of an anticlimax, for independent observers had been busy studying the local floras of Europe and the British Isles with a fresh eye, and as knowledge grew and printing developed, herbals with enlightened text and excellent illustrations appeared. Even so, authors introduced many quotations from Dioscorides. An Italian physician, Pierandrea Mattioli, wrote a 'commentary' which, combined with his own knowledge, became a monumental and much sought work. Mattioli became physician to the Austrian Emperor Maximilian II and later died of the plague in 1577.

To return to the rational if less picturesque path of botanical learning laid by the aforementioned philosopher and his teacher, Aristotle: Theophrastus was born on the island of Lesbos (Mytilene) off the coast of Turkey, a locality then celebrated for its gardens, grapes and wine, and earlier for a circle of poets called the Aeolian School, of which Sappho, a noblewoman of the island, was the leading light. Admired by Aristotle and Plato, she was named 'the Tenth Muse.' The Aeolians were one of three major elements of the Greek race, the other two being the Dorian and Ionian, who gave their names to the capitals of Greek architecture. In this environment the scientific studies of Theophrastus naturally turned to plants. The Aristotelian approach was one of patient and logical inquiry into the nature of living organisms. Ultimately this method, as exemplified in Theophrastus' *Enquiry into Plants*, led to modern identification.

Many ferns in this book were named by Karl von Linné (1707—1778), the famous Swedish botanist who originated the binomial system used in modern scientific naming of plants. The first name stands for the genus (genera in the plural). The second denotes the species. An initial L follows the name given to each plant by Linné, who is generally known by the Latinized form, Linnaeus. Some ferns were placed by later botanists in different genera, or their specific names were changed. The abbreviated name of the authority follows the plant name. The scientific names at each heading in this book are the ones presently preferred by fern experts.

In the three sections denoted on the title page, the plants are arranged in alphabetical order by genera. Others of peculiar interest may be found in the appendix, in the same arrangement.

For inspiration, advice, and aid in the study of specimens I am grateful to Dr. Rolla M. Tryon and Dr. Alice F. Tryon, Gray Herbarium, Harvard University; and to Dr. David Barrington, Pringle Herbarium, University of Vermont.

Elfriede Abbe

Manchester Center, Vermont
June 27, 1980

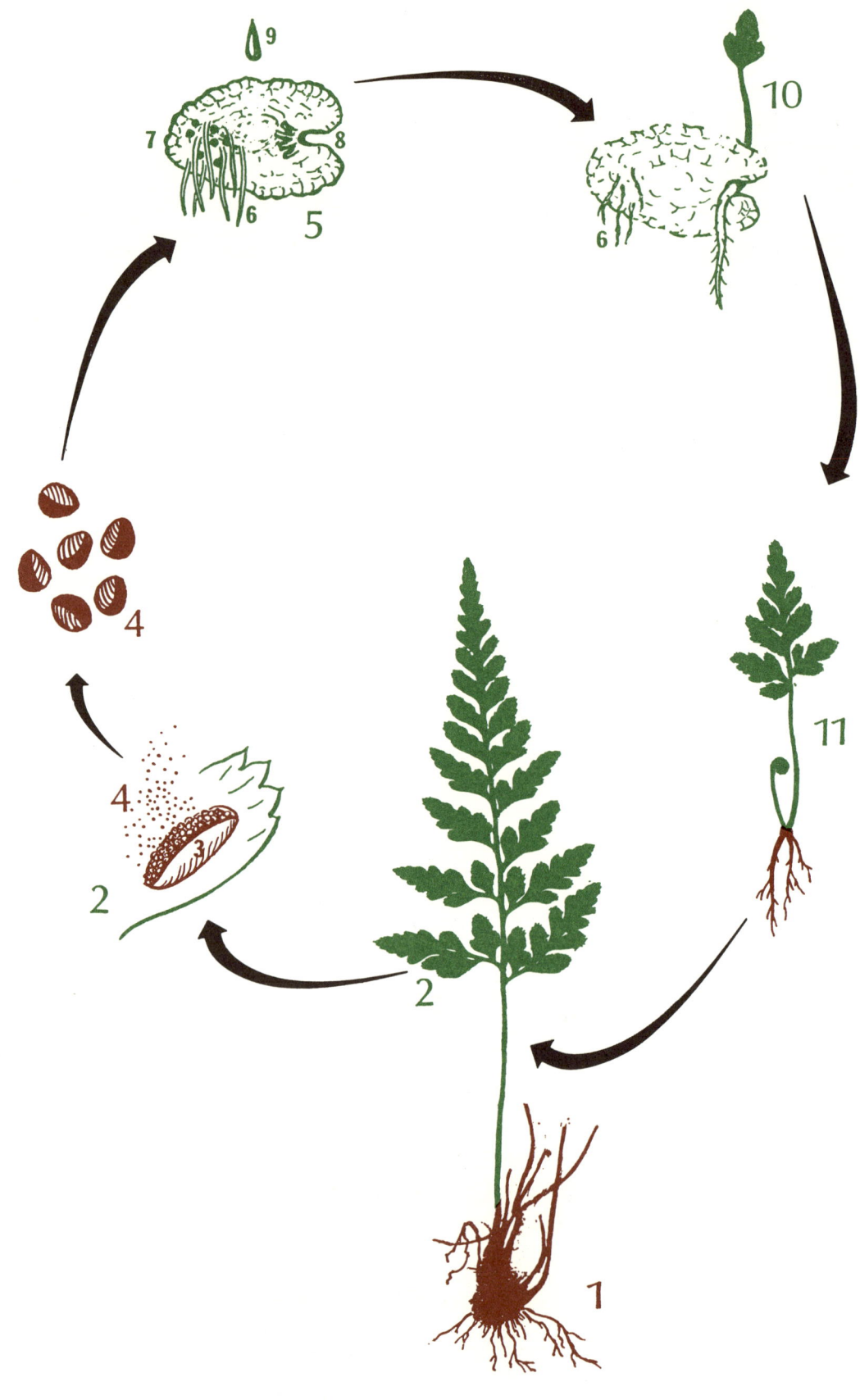

Life Cycle of Fern

The Fern Plant

In the evolution of plants, the first to have roots and leaves were the ferns. The ancestors of modern ferns appeared in the Devonian age, 280—325 million years ago, along with the forebears of coniferous trees and the horsetails and club mosses. They flourished, and about 75 million years later, in the Mesozoic age, a period of nearly uniform warmth and moisture over the entire earth, they achieved the magnitude of trees, some of the club mosses reaching a height of one hundred feet. Woody fibers and a system of vessels gave the ferns physical strength and a circulatory system for conveying moisture and nourishment throughout the plant.

Unique in the vegetable kingdom for the characteristic structure of their leaves and more particularly for their peculiar manner of reproduction, the ferns form a separate order among plants, called the Filicales. With the allied orders of the Equisetales (Horsetails) and the Lycopodiales (club mosses), they constitute the division of spore-bearing plants called Pteridophytes, the highest division in the plant kingdom next to the seed-bearing plants.

Fern leaves rise directly from the rhizome on more or less elongated stalks or petioles. The rhizome or rootstock is actually a stem which may grow above or below ground. It is the trunk of the tree ferns. The rhizome produces roots which find water and nutrients for the plant. The whole fern including its rhizome is called the *sporophyte*, and is shown as 1 in the diagram opposite.

The venation of most ferns is remarkable for its regularity, each vein dividing and sub-dividing into two equal forks, a system known as dichotomy. Some ferns have unforked veins. Only a few species have netted veins. In the presence of forked veins there is a tendency for the plant to proliferate foliage beyond normal growth. Horticulturists of the eighteenth and nineteenth centuries observed this and developed innumerable fantastic forms with crested, tufted, fringed, and scalloped leaves.

On their undersides fern leaves bear clusters of minute capsules which are called *sporangia*, at 2 on the diagram above. Each sporangium contains numerous spores (4). The clusters are called *sori* (singular *sorus*), from Greek meaning heap, and are variously arranged on veins and margins of leaflets in the different species. In a few cases the sori are terminal and are mingled with minute leaflets, as in *Osmunda*; or are terminal on naked stems as in the rattlesnake fern. The membrane which covers the spore clusters of many species is called an *indusium*, from the Latin for tunic (3).

The spores are microscopic, appearing to the unaided eye somewhat like dust or finely ground pepper. They are shed mostly in June and July. Falling on damp soil, the one-celled spore enlarges, and divides to form other cells. Gradually this aggregate of cells assumes a flat heart-shaped or kidney-shaped form about a quarter of an inch in diameter, somewhat resembling a lichen; this new green plant is an intermediary vegetative organism in the life cycle of the fern and is called a *prothallium*, meaning 'first growth,' or a *gametophyte*, 'reproductive plant.' It is 5 on the diagram.

From the underside of the gametophyte, rhizoids sprout (6) which reach into the ground and function as roots. Near them the *antheridium* develops, which produces male sperm cells (7); at the opposite side, close to the notch of the gametophyte is the *archegonium* (8) where an egg cell develops. For the sperm cells to swim to the archegonium to fertilize the egg a film or drop of water (9) is needed. This dependency on water in the life cycle accounts for the preference of ferns for moist locations. After fertilization the cell divides

to make more cells. These form the first root and bud of a new sporophyte, or spore-bearing plant (10), which at first bears no resemblance to its mature self, as seedlings of higher plants look unlike the parents. The prothallium, having served its purpose, withers and dies, and the new plant (11) grows by its own rhizome and rootlets. Ferns reach the 'fiddlehead' or 'crozier' stage of uncoiling fronds, shown on page 1, during April and May. The fully developed leaves bear the spores, which are shed in midsummer, and so the fern's life cycle continues.

Although some ferns have simple leaves (1, opposite page), compound leaves characterize the majority. An intermediate form, in which the blade is dissected almost to the midrib, is called a pinnatifid leaf (2, opposite). Terms and synonyms for the component parts of a compound leaf are given in the diagram above. In a decompound leaf, the leaflets are divided into subleaflets; this form is also called bipinnate or twice-cut. Further subdivisions are usually called lobes. Pinna (plural *pinnae*) is from Latin for feather.

The sori, sporangia and spores are of major importance in determining the identification of species of ferns. In modern scientific research the electron microscope reveals a remarkable diversity and complexity in the form and structure of spores, which were imperfectly known before the invention of this powerful modern instrument, capable of showing plant tissues measured in molecules.

A sorus (cluster of sporangia) on the fern blade is often called a fruit-dot. The term, although popular, is neither accurate nor appropriate. *Fruit* in a biological sense does not exist in ferns, and *dot* is an indefinite word, which, if used to indicate the shape and general distribution or arrangement of sori, is applicable in only a limited number of cases.

I
The Ferns

Adiantum capillus-veneris

Adiantum capillus-veneris L.

Southern Maidenhair, Venus' Hair

Fr.: Capillaire commun, Capillaire de Montpellier, Cheveu-de-Vénus
Ger.: Frauenhaar, Wasserkoriander   It.: Capelvenere

¶ *Description*
Leaves lanceolate to ovate, evergreen, to 12 inches in length. Near the base usually bipinnate, one-pinnate above. Stalk and axis thin, decumbent, wiry, very dark and shiny. Leaflets delicate in substance, fan-shaped, wedge-shaped or rhombic, upper margins lobed or finely toothed. Sori covered by inrolled upper margins of leaflets, as shown enlarged in the plate opposite. Rhizome long, horizontal. In his 'Herball of Large Extent,' the *Theatrum Botanicum*, published in London in 1640, John Parkinson says, '... fine leaves without order on both sides one above another, somewhat like unto the lower leaves of Coriander, or like the leaves of Anise.' Dioscorides was actually the first to make the comparison. Coriander and anise are condimental herbs of the Carrot Family and are native to the Mediterranean. A common wild plant, the tall meadow rue, and several related species of the genus *Thalictrum* of the Crowfoot Family are often mistaken for southern maidenhair because of a superficial similarity of the leaves to the leaflets of the fern, although there is no substantial resemblance between the two. Maidenhair should not be confused with mermaid's-hair, an alga with thin bright green filaments, growing in rocky streams and waterfalls.

¶ *Habitat and Distribution*
In warm-temperate to tropical regions of both hemispheres. In Europe rare, but found in warm damp regions of southern and western Europe, ranging northward up the Atlantic coast to the British Isles, where it is found in Cornwall and Devon and along the west coast of Ireland. In America the fern appears in scattered areas from Virginia and Kentucky southward to tropical zones; in the West, from Utah

and California to Missouri. Inhabits the margins of waterfalls, rocky streams, springs and damp caves in limestone regions.

¶ *Culture* Suitable for greenhouse or as a house plant in the north. The root system requires continuous moisture and a mildly alkaline condition which may be maintained by frequently repotting the fern in light loamy soil to which ground limestone has been added. Limestone pebbles should be used for drainage at the bottom. The plant should be kept away from dry heat and direct sunlight.

¶ *History* This is the white *adianton* of Theophrastus. John Gerard, who was Master of the Barber-Surgeons' Company in London and superintendent of the gardens of Lord Burghley, Queen Elizabeth's secretary of state, says in his *Herball* that Apuleius in the fifth century was the first to use the name *capillus-veneris*, and he adds, 'It is called Adianton because the leafe is never wet, for it casteth off water that falleth thereon, or being covered in water, it remaineth still as if it were dry.' The name is from Greek *a*, (without) and *diainem* (to wet). The plant grows principally along margins of watercourses, but a waxy epidermis causes the leaflets to shed water and stay dry. In the twelfth-century Bodley Manuscript of Apuleius it is stated that the Greeks called the fern *capillus veneris*, which is the Latin rendering of 'Aphrodite's hair.' There is said to have been a tradition in Greek mythology that when Aphrodite rose from her birth in the sea her hair was dry. Ocean winds blew upon her, and Botticelli welcomes her with the landward breath of the West Winds.

How a small fern inherited this combination of names is lost in legends; Sir Thomas Browne (1605—1682) gave form to the last phase of romantic fantasy when he used the word 'capillary' as a noun, meaning a fern that grows like tufts of hair. John Ray in *Historia Plantarum* (1686) and Hermann Boerhaave (1668—1738), professor of botany at the University of Leiden, both used the term to describe ferns. Five ferns enjoyed the distinction of being 'the capillary herbs.' They were the southern maidenhair, common polypody, common spleenwort, wall rue and hart's-tongue fern.

¶ *Use* Dioscorides recommends *adianton* as a remedy for asthma, the stone, splenetic disorders, snakebite. As a cure for falling hair he prescribes a mixture of maidenhair, *ladanum* (cistus gum from the plant *Cistus creticus*), *oesypus* (wool fat or lanolin), and myrtle oil; or a decoction of maidenhair rubbed on with lye and wine. Apuleius recommends Venus'-hair for sciatica.

Medieval and Renaissance herbalists adopted the uses of maidenhair from ancient writers. Gerard says it is a remedy for the stone, congestions of the chest and lungs, 'and maketh the haire or beard to grow.' John Parkinson says, 'Maidenhaire is of singular good use against diseases of the Breast, the Liver and Reines. The decoction of the herbe drunk helpeth Cough, the yellow Iaundies, diseases of the Spleene. The herbe boiled in oyle of Camomill dissolveth knots, alayeth swellings . . . The Lye made thereof is singular good to cleanse the haire, stayeth the falling and shedding of the haire, and causeth them to grow thicke, faire and well coloured, for which purpose some boyle it in wine, putting some smalledge seede thereto, and afterwards some oyle.' Seed of wild celery or smallage contains aromatic oils formerly used in medical preparations.

Lye is water impregnated with alkaline salts of wood ashes by the process of leaching, the ashes being placed in a barrel or similar container with a sieve or network of reeds or twigs for drainage at the bottom, and water poured on them to percolate through the mass. The resulting liquid is lye, mostly potash. Primitive soap was a mixture of lye, fats and oils, boiled together in a caldron. Fern ashes are strong in alkali but are purer and milder than wood ashes. Tincture of green soap, an officinal preparation of potash and linseed oil, is used as a medicinal shampoo and in the treatment of skin diseases. A tincture is a solution of macerated material in alcohol; wine provided the alcohol in earlier preparations.

The ancient reputation of maidenhair is affirmed by the French name, *Capillaire de Montpellier*. A distinguished school of medicine was founded in the twelfth century at Montpellier, a town in southern France near Marseille.

The school of botany of the university was highly esteemed, and many distinguished herbalists studied there, including De l'Obel, Clusius, Bauhinus, and d'Aléchamps. A cough medicine which originated in Montpellier was called Syrup of Capillaire. It is made by pouring 1 pint of boiling water on 1 ounce of fresh maidenhair fronds. The infusion, after steeping 6 to 7 hours, is strained and mixed with 4 ounces of orange-flower water. Two lbs. honey are warmed over a low fire, and the liquid slowly stirred in. In later times licorice root was included as part of the infusion, and sugar syrup was used instead of honey.

John Sowerby in *The Ferns of Great Britain*, 1855, says that in the South Isles of Aran off the coast of Galway, Ireland, where the maidenhair grows profusely in the fissures of the limestone rocks, the people use a decoction of the fronds as a substitute for tea. A strong decoction acts as an emetic. The planet Venus must not have been *in the ascendant* when Culpeper observed, 'All the Maidenhairs should be used green and in conjunction with other ingredients because their virtues are weak.'

Adiantum pedatum

Adiantum pedatum L.

Maidenhair Fern, Northern Maidenhair, True Maidenhair

¶ *Description* Outline of the leaf horseshoe-shaped or nearly circular. Stalks slender, wiry, black or brown, rising erect from along the creeping rootstock. Plant 10 to 20 inches high. Leaflets mostly oblong. Subleaflets fan-shaped to rhomboid; terminal ones rounded-triangular. Upper margins of the subleaflets bear the sori, which are covered by the infolded margins of the subleaflet. The pattern of the leaf is reminiscent of the shape of an animal's paw, and the branching of the axis suggests the anatomical subdivisions of the bones of the foot; the similarity gave rise to the specific name *pedatum*, Latin meaning footed.

¶ *Habitat and Distribution* The northern maidenhair is a native of North America. It ranges from Nova Scotia to British Columbia, and south to Georgia and Arkansas, preferring damp shady places, rocky banks and rich forest humus. The northern maidenhair is a plant of limestone regions. Also found in eastern Asia.

¶ *Culture* In rich, moist, well-drained neutral soil *A. pedatum* does well as a garden plant. It spreads readily and forms an excellent and decorative ground cover in moist shady areas or on the north side of buildings. The rhizomes must be kept wet during and immediately after transplanting, until the fern is established. This species is also easily started from spores.

¶ *History* Jacques Philippe Cornut, in his *Canadensium Plantarum* of 1635, calls this fern *Adiantum Americanum* and *A. Canadensi*. Parkinson in *Theatrum Botanicum*, 1640, says, 'Adianthum fruticosum Americanum, Forraine or strange Maidenhaire groweth up like unto a Ferne, with a slender blackish browne stalk branched forth into others whereabouts on each side stand from 12 to 20 fresh greene leaves... One very like unto this Mr. Iohn Tradescant brought out of Virginia.' Tradescant was a famous plant collector and fancier and a friend of Parkinson. The latter was apothecary to Charles I, and Tradescant held the position of supervisor of the royal gardens.

In John Bartram's *Observations in his Travels from Pennsylvania to Onondago, Oswego*, etc., published in London in 1751, appears this report: 'July 15. We set out a N. E. course, and passed by very thick and tall timber of beach, chesnut, linden, ash, sugar-maple... some white pine, with ginseng and maidenhair. July 18. This morning we sent an Indian with a string of Wampum to Onondago to acquaint them with our coming . . . This day our general course was N. and N. W. . . . tall timber oak, birch, . . . maidenhair in abundance.' Onondago was the region south of Oneida Lake in New York State. The ginseng plant was valued by the Indians for its tonic properties; when the French Jesuits in Canada learned of the plant from the natives, they began to export large quantities of the root to China, where there was a great demand for ginseng as a cure-all.

¶ Use

A. pedatum was used for the same purposes as *A. capillus-veneris* but was more highly esteemed for its supposed superior strength. In *New England's Rarities*, 1672, John Josselyn calls the plant 'Maiden Hair, or *Capillus veneris verus*, which ordinarily is half a Yard in height. The Apothecaries for shame now will substitute *Wall-Rue* no more for *Maiden Hair*, since it grows in abundance in *New-England*.'

Per Kalm, the Swedish traveller who collected many American plants for Linnaeus, says in the *Travels in North America* (1748—1750), 'The plant which throughout Canada bears the name of *Herba capillaris* is one of those with which a great trade is carried on in Canada. Several people in *Albany* and *Canada* assured me that its leaves were very much used instead of tea, in consumptions, coughs, and all kinds of pectoral diseases. This they have learnt from the Indians. This American Maiden-hair is reckoned preferable to that which we have in Europe; and therefore they send a great quantity of it to *France* every year. The price is different, and regulated according to the goodness of the plant, the care in preparing it, and the quantity to be got.'

A. pedatum has been useful to the Indians not only medicinally, but also for the decorative character of the dark shiny stems in basketry.

Asplenium adiantum-nigrum

Asplenium adiantum-nigrum L.

Common Black Maidenhair, Black Oak Fern, Black Spleenwort, Male Black Fern, Moss Fern

Fr.: Capillaire noire, Doradille noire   It.: Asplenio
Ger.: Schwarzer Milzfarn, Streifenfarn

¶ *Description* Leaves elongate-triangular, somewhat leathery, evergreen. Stalks dark and shiny. Leaflets elongate-triangular, alternate, ascending, irregularly dentate, lowest pair longest. Plant grows in tufts from the rhizomes and may reach a height of 2 feet in a moist shady environment; average height about 6 inches. Sori (shown enlarged in opposite plate) are elongate, with thin indusium. Plant usually dark green.

¶ *Habitat and Distribution* Common in Europe and the British Isles, southern Scandinavia, the Netherlands, Belgium, France, and southward to the Mediterranean region, the Balkans, Caucasus and North Africa. On gravelly siliceous slopes and shady banks in mountain regions; also among boulders, in rock crevices and on sandstone ledges, old walls and roots of trees.

¶ *Culture* Well adapted to the rock garden if set in rich sandy loam over a well drained sandstone base. Not a plant of limestone areas.

¶ *History* The name *Asplenium* is derived from the Greek word for spleen. Several ferns were so named by the ancients as cures for enlarged spleen and for liver and kidney complaints. In the German name of the fern, *milz* means spleen and is of the same origin as the old English word milt, derived from ancient Icelandic and still used in some localities.

We have an American fern which is sometimes called black maidenhair, but it is not the herb *Asplenium adiantum-nigrum*. Our species is more often called ebony spleenwort, and its scientific name is *A. platyneuron* (L.) Oakes.

Black maidenhair formed the basis of a cough medicine for asthma ¶ *Use* and lung diseases, the mucilaginous qualities of the plant being concentrated by decoction. A mild tonic property, common to several species of ferns, has given the black maidenhair repute as an aperient.

Gerard says that 'Male blacke Ferne is like in faculty to Trichomanes or English Maidenhaire.' (See *Asplenium trichomanes*.) Parkinson in *Theatrum Botanicum* gives the fern the name *Adianthum nigrum vulgare* and observes, '... the virtues the same as Dryopteris of which Mattiolus saith that the roote in powder with a little salt and Branne is given to Horses for the wormes... species of Dryopteris are used in the Apothecaries shops of divers countries for *Adianthum album* and *nigrum*.' (See *Dryopteris filix-mas*.)

Asplenium ruta-muraria

Asplenium ruta-muraria L.

Wall rue, Rue maidenhair, Rue fern, Rue spleenwort, Salvis vitae, Stone fern, Taintwort, Tentwort, Wall-pie, White maidenhair

Fr.: Rue de muraille, Doradille de muraille   It.: Ruta di muro
Ger.: Mauerraute, Mauerstreifenfarn

¶ *Description*   The branching habit and the shape and color of the leaflets give this fern considerable resemblance to garden rue. Stalks slender, growing in a cluster from the short rootstock, which is covered with pointed brown scales. Whole plant about 5 inches high. Leaves ovate, somewhat leathery. Leaflets ovate, on stalks spaced alternately along the axis. Triangular, rhombic or unsymmetrically fan-shaped subleaflets with upper margins irregularly incised. Under favorable growing conditions the sori may completely cover the undersides of the leaflets; otherwise widely spaced on the veins as illustrated opposite. The entire plant is a pale blue-green like the color of rue; the similarity gave the fern its common name.

¶ *Habitat and Distribution*   In limestone regions, on rocks and cliffs, in shale or in moist crevices. As its name indicates, the fern also frequently grows on broken masonry, the north side of old walls, church towers, bridges and ruins. Wall rue requires shade or partial shade and neutral soil. Very common in Europe and the British Isles. Also ranges from the Near East to the Himalayas. Rare in the United States, where it ranges from northern Vermont to northern Michigan and southward to Virginia and Missouri.

¶ *Culture* Not easy to cultivate, but can be grown in the rock garden if the rootstocks are kept continuously moist after transplanting and during period of first growth until roots are established. Requires calcareous neutral soil. British recommendations emphasize using old mortar in the soil.

¶ *History* Wall rue has long been considered a specific remedy for scrofula, an infection characterized by swelling of the glands of the neck and by a general physical debility. Children are especially subject to the disease. Sometimes called 'the taint,' scrofula was also commonly known as 'the king's evil,' because it was thought to be curable by the touch of the king's hand. Edward the Confessor (1042—1066) was the first sovereign to whom this power was attributed. The belief persisted in England through the seventeenth century. By popular demand, Charles II began to 'touch for the evil' in 1660, the year of the Restoration.

Wall rue derived its reputation from rue because of the outward resemblance of the two plants. Pliny (A.D. 23—79), the Roman compiler of the *Historia Naturalis*, quotes a Greek authority as saying there are two kinds of rue, differing somewhat in form and color of leaf.

¶ *Use* Pliny enlarges on rue as a virtual cure-all, specially mentioning its application against scrofula. In the words of Philemon Holland, an English translator of Pliny's work (1601), 'The inunction thereof with Allum and Honie cleanseth leprosie. Likewise with Nightshade, Hogs grease and Buls tallow it dispatcheth the Kings evill.' Wall rue was administered in the same way, as a poultice made from the macerated fronds mixed with lard.

The dairymaids of the Austrian and Tirolian Alps daily feed a cluster of wall rue and other herbs to the cows as a charm against bewitchment. Many cases of sickness of cattle, theft and souring of milk have been attributed to witchcraft, especially during the seventeenth-century witch persecutions.

Asplenium trichomanes

Asplenium trichomanes L.

English maidenhair, Black-stemmed spleenwort, Common maidenhair, Common spleenwort, Dwarf spleenwort, Maidenhair spleenwort, Baby fern, Wall spleenwort, Waterwort fern

Fr.: Capillaire, Doradille, Polytric officinal It.: Erba ruginina
Ger. Brauner Milzfarn, Haar Milzfarn, Steinfeder, Widertod

¶ *Description*

Average height of plant 5 inches. Sterile fronds many, narrow and tapering, decumbent, clustered, dark green with brown stem, persistent. Leaflets in approximately 15 to 20 pairs, oval or nearly round, upper margins wavy, lobed or eared. Fertile leaves erect, lighter green. Sori with indusium, oval or oblong, located on lateral veins of leaflets, often covering entire undersurface. Rootstock short, erect, chaffy, with brown scales.

In Gerard's description, 'English Maiden-haire hath long leaves of a darke green colour, consisting of very many small round leaves set upon a middle rib, of a shining blacke colour, dashed on the nether side with small rough markes or speckes, of an overworne colour.' These are the dark brown sori. The 'female English Maiden-haire' to which Gerard refers is probably the sterile portion of the plant, for he says the leaflets lack 'the spots or markes' of the other.

¶ *Habitat and Distribution*

In Eurasia, North and South America, Australia. Common in southern Europe, also in the Caucasus, the Urals and the Himalayas. In moss on limestone rocks and cliffs, usually in moist shady locations in moderately acid soil. In North America ranges from Alaska, Minnesota and Nova Scotia south to Arizona and Georgia. Found in Vermont at elevations up to 2500 feet. Sowerby says that except for the polypody, this is probably the most abundant of the rock-loving ferns in the

British Isles. Parkinson says it grows 'upon old stone walls in the west parts of England and Wales, in Kent and divers other places.'

¶ *Culture* If the fern is transplanted from the wild, the safest method is to take the plant with the rock on which it is growing, to avoid damaging the roots. If not movable, the rock may be carefully chipped. Pot-grown plants may be had from nurseries. English maidenhair does best in moist well-drained neutral soil in limestone crevices. British authorities advise mixing decayed mortar with the soil to provide the potash and lime that are necessary for most ferns.

¶ *History* The name *trichomanes* is Greek, meaning a mania for hair, and it refers to the manner of growth of the fronds. (See 'capillary herbs,' page 8.) Theophrastus calls this fern black *adianton*.

¶ *Use* *A. trichomanes* was formerly the principal officinal ingredient of a French elixir called polytrichon or kalliphyllon, the dose 1 ounce of fronds infused in a pint of boiling water, afterward mixed with honey. An elixir is an aromatic, sweetened preparation containing medicinal substances and sometimes spirits. English maidenhair is mucilaginous, sweetish and astringent, and has a sweet odor when dry. For its expectorant properties, the fern has often been substituted for southern maidenhair in Syrup of Capillaire. The astringency of the plant has led to some use as a tea in parts of Ireland. The peasants of the European Alps consider the fern a charm against witchcraft and other evils.

Athyrium filix-femina

Athyrium filix-femina (L.) Roth

Lady fern

Fr.: Fougère femelle   Ger.: Weiblicher Waldfarn   It.: Felce femina

¶ *Description* The leaves grow in a circular tuft and may reach 30 inches or more in height; outline broadly lanceolate with acuminate decumbent tips. Fronds divided into numerous pairs of somewhat upwardly ascending leaflets. Subleaflets in several pairs, margins deeply incised. Stalk shorter than blade. Sori and indusia short, curved. In the form of the fronds this fern is so variable as to be baffling to amateur attempts at identification. Rootstock slender, mostly ascending.

¶ *Habitat and Distribution* Common in America, Europe and the British Isles. Grows in woods, rocky areas, and edges of clearings, in moist neutral soil and full or partial shade. Likes moist beech woods. More frequent in lowlands than at higher elevations. In *Waverley* Sir Walter Scott hies away to the habitat of the lady fern:

> Where the copsewood is the greenest,
> Where the fountains glisten sheenest,
> Where the lady fern grows strongest,
> Where the morning dew lies longest,
> Where the black-cock sweetest sips it,
> Where the fairy latest trips it:
>   Hie to haunts right seldom seen,
>   Lovely, lonesome, cool, and green.

The blackcock is a common European grouse with black feathers and lyre-shaped tail.

¶ Culture
Easily grown in the garden in moist shady areas. Does well in most kinds of neutral or moderately acid soil. Excellent and decorative ground cover for banks and bare shady spots.

¶ History
The generic name *Athyrium* is from the Greek *athyros*, meaning open; the indusium opens and turns completely away from the surface of the leaf when the spores are ripe. The name *filix-femina* originates from the Greek and Roman concept of sex. The ancients were unenlightened about the reproductive processes of plants. In comparing two plants closely resembling each other, the ancients called the larger, coarser or more vigorous one the male, and the smaller, more delicate plant the female. For centuries there was confusion of male fern, female fern and brake, because of a superficial resemblance of the foliage of the three ferns. Theophrastus called one of these *Thelypteris* (female *pteris*), which Romans translated into *filix femina*. William Turner, a physician and clergyman called the 'father of British botany,' sums up the commonly accepted identifications in this passage of 1548, but perpetuates the confusion of the lady fern with brake: 'Filix is called in greeke Pteris, in english a Ferne or a brake. There are two kindes of brakes. The one kynde is called in latin Filix mascula and in greeke Pteris. The seconde kynde is called in greeke Thelypteris, in latine Filix femina, that is the commune fern or brake, which the Northerne men cal a bracon.'

Bock gave the lady fern a sensible name, *filix vulgaris*, Waldfarn (common forest fern), but the ancient classical name persisted. The courtly eighteenth century and the proper nineteenth changed the female to lady.

¶ Use
Has the same anthelmintic properties as *Dryopteris filix-mas* but to a lesser degree; the rhizomes of both herbs are frequently mingled in pharmaceutical preparations. The American Indians have used the plant as a diuretic and the rootstock as a tonic or bitter, similar to European use of male fern for a stomachic. Lady fern is very common in Ireland and is used there for packing fruit and fish.

Botrychium lunaria

Botrychium lunaria (L.) Sw.

Moonwort, Moonwort grape fern, Lunary, Blasting-root, Spring-wurzel, Unshoe-the-horse

Fr.: Botrique a croissants, Botrique lunaire   It.: Erba lunaria
Ger.: Gemeine Mondraute

¶ *Description* Plant fleshy. Single frond oblong; three to six pairs of leaflets in opposite pairs, fan-shaped or half-moon shaped, often overlapping, margins wavy, sometimes notched. Average height of sporophyll stalk 4 inches; short terminal branches bear sporangia. Bud for the following year enclosed in base of main stem.

¶ *Habitat and Distribution* Dry grassy meadows and hillsides in limestone regions. Culpeper remarks, 'It groweth where there is much grass, for therein it delighteth to grow.' A rare plant of Europe and the British Isles. Rare in the United States, in scattered localities from Newfoundland southward to Connecticut and central New York; in the West, British Columbia and the Rocky Mountains.

¶ *History* *Botrychium* is from the Greek *botrys*, a bunch of grapes, and refers to the cluster of sporangia. According to the Doctrine of Signatures the crescent-shaped leaflets indicate that the plant is governed by the moon. Alchemists used the moonwort to convert quicksilver (mercury) into silver, the latter metal being under the influence of Luna (the moon) because of her silvery color. The disillusioned Canon's Yoeman of Chaucer's *Canterbury Tales* mentions among the ingredients of his master's pursuit:

> Waters rubifying and bulles galle,
> Arsenik, sal armoniak, and brimstoon;
> And herbes coude I telle eek many oon,
> As egremoin, valerian, and lunarie

Sir Thomas Browne cites the belief that '*ferrum equinum* (Latin meaning horse iron, another name for moonwort) hath a virtue attractive of Iron, a power to break lockes, and draw off the shoes of a horse that passeth over it . . . Which strange and magicall conceit seemes unto me to have no deeper root in reason than the figure of its seed, for therein indeed it somewhat resembles an horseshoe, and which notwithstanding Baptista Porta hath too low a signation, and raised the same unto a Lunatic representation.' This quotation is from *Pseudodoxia epidemica*, 1646.

Giambattista Porta was a learned Italian astrologer who devoted his major work, the *Phytognomonica* (1588), to the Doctrine of Signatures and had great influence in spreading the Doctrine. One of his illustrations shows the moon accompanied by moonwort and half-moon-shaped seed pods of two other plants. Sir Thomas should have looked again.

William Cole in 1656 had, of course, furthered the notion that ¶ *Use* 'Moonwort will open the locks wherewith dwelling-houses are made fast, if it be put into the keyhole.' 'This,' says Gerard, 'some laugh to scorn, and those no small fools neither; but country people that I know call it Unshoe the Horse. Besides I have heard commanders say, that on White Down in Devonshire near Tiverton there were found thirty horse-shoes, pulled off from the feet of the Earl of Essex's horses, many of them being but newly shod, and no reason known which caused much admiration . . .' Elsewhere, however, he says of witches' and alchemists' claims for the plant, '. . . in truth they are all but drowsie dreames and illusions.'

Various nerve and brain disorders, such as brain tumor, epilepsy and sleepwalking, were supposed to be related to the phases of the moon because of the victims' periodic seizures. Lunacy became the name for irrational behavior. Moonwort had to be a cure, provided that it was gathered by the light of the full moon. Peasants of the Alps claim the frond turns with the moon, and that the fern causes reduction of milk and loss of calves in their herds.

As an ancient vulnerary, the herb was macerated in oil or melted suet to make a salve. According to Culpeper, 'The leaves boiled in red wine and drank . . . stayeth bleeding, vomiting and other fluxes. It helpeth all blows and bruises, and to consolidate all fractures and dislocations.'

Botrychium virginianum

Botrychium virginianum (L.) Sw.

Hemlock-leaved moonwort, Rattlesnake fern, Snake brake, Virginia grape fern

¶ *Description* Including the mature sporophyll stalk, total height of plant may reach two feet. Stems fleshy. Leaf triangular, wider than long, ternately divided. Leaflets opposite, mostly pointed. Sporophyll stalk with six or more branches bearing the sporangia, shown enlarged opposite. Base of mature plant encloses buds of the following seasons.

¶ *Habitat and Distribution* A native American plant found in damp shady woodlands in rich humus, neutral to mildly acid. Common from southern Canada southward to Florida and California.

¶ *Culture* In taking a plant from the wild, include as much of the native soil as possible around the roots. The constitution of woodland soils is subtle and not easily duplicated in the garden. *B. virginianum* is attractive to slugs.

¶ *History* During the period of colonization large quantities of ginseng were exported from this continent to China, where the plant has always been highly esteemed in medicine. A large population of rattlesnake fern was eagerly sought by settlers and explorers, for the fern was considered an indicator of the presence of ginseng, especially in the mountains of Kentucky and Tennessee, since the two plants sometimes grow in a common environment.

¶ *Use* The American Indians have used the plant for treating wounds and as a specific remedy for the bites of poisonous snakes; in the latter case, a poultice of the fresh macerated root is applied. Josselyn (1672) informs us that the Indian method of treating wounds was by 'annointing the wound first with Racoons greese or Wild-Cats greese, and strewing upon it the powder of the Roots.' To keep snakes out of the wigwams, the women made a decoction of the root, with which they sprinkled the ground.

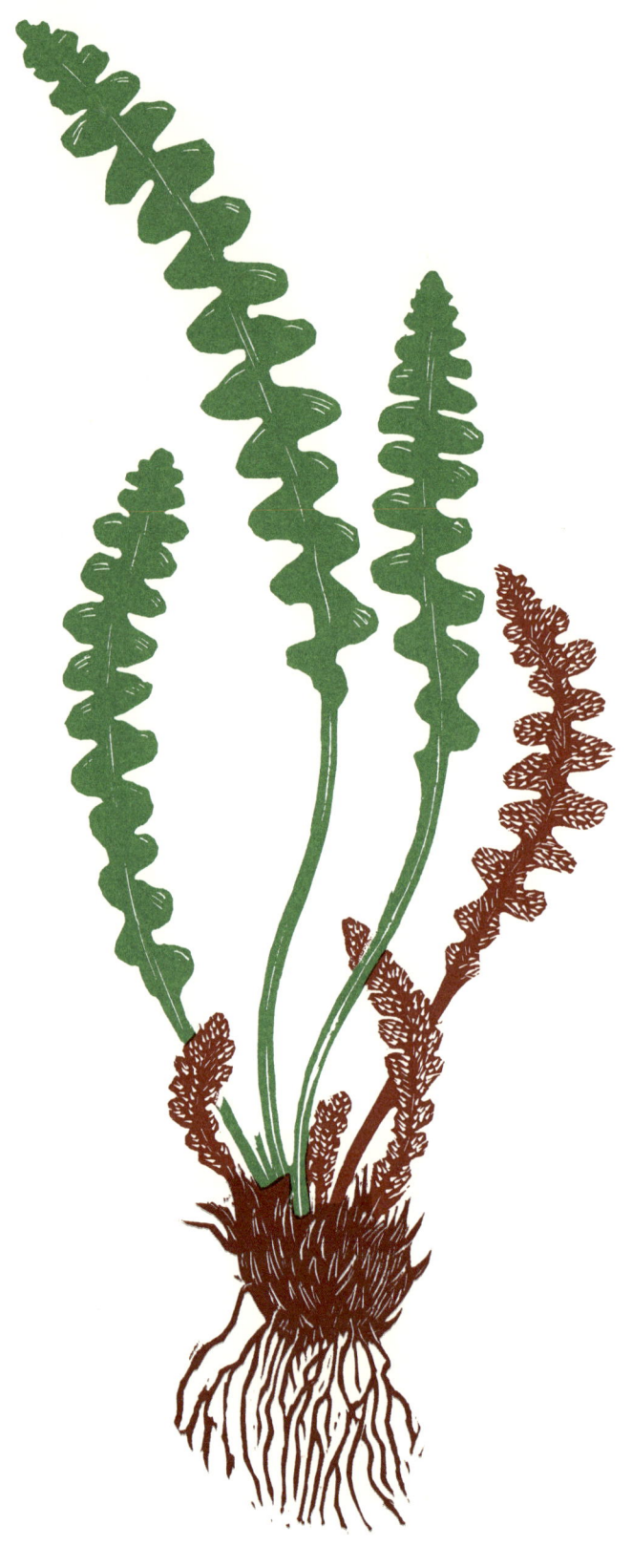

Ceterach officinarum

Ceterach officinarum Lam. & DC.

Ceterach, Common spleenwort, Finger fern, Miltwaste, Scaly spleenwort

Fr.: Cétérach, Doradille   It.: Cetracca, Erba ruggine
Ger.: Schriftfarn, Schuppenfarn, Spreuschuppiger Milzfarn

Leaves linear-lanceolate, dull gray-green, leathery, deeply cut ¶ *Description* into alternate obtuse segments. Average height of plant about 5 inches. Under surface of mature leaves entirely covered with brown, chaffy, pointed, overlapping scales. Sori oblong, hidden by the scales.

In Culpeper's words, the herb 'from a black, thready and bushy root, sendeth forth many long single leaves, cut in on both sides into round dents almost to the middle . . . each division being not always set opposite unto the other . . . a dark yellowish roughness on the back, folding or rolling itself inward at the first springing up.'

Indigenous to the Mediterranean region and the islands ¶ *Habitat and* of Madeira, the Azores and the Canaries. Ranges north up *Distribution* the eastern North Atlantic coast. Found in western England. Common in Ireland. On the Continent extends to Asia Minor and Tibet. Inhabits limestone rocks and cliffs, walls and ruins in old mortar. Also grows in damp but well drained soil of a calcareous nature.

This herb is considered to be the true spleenwort of the ancients ¶ *History* and is the plant which they believed responsible for wasting away the spleens of the swine that browsed on the fern on the island of Crete. Dioscorides describes the leaves as 'beneath yellowish and rough, above green,' recommends the plant for splenetic ailments and the stone, and in the words of his translator, Sir John Goodyer (1655), advises that the herb 'be digged up when ye night is moonless.'

In ancient rituals when animals were slaughtered as sacrifices to the gods, the various entrails were burned separately, and their condition was noted by the celebrants as auguries; so the state of the swines' spleens did not pass unobserved. The Greek festival of the Olympic Games was one of the events at which pigs were sacrificed by the participants.

The name ceterach is from medieval Greek *kitarak* and Latin *ceterah*, corruptions of *chetherak*, the name given to the herb by ancient Arabian and Persian physicians. *Officinarum* from Latin *officina*, laboratory or apothecary's shop and later office, is applied in pharmacy to preparations made according to recognized prescriptions, and in botanical connotations denotes a plant used in medicine. This specific name indicates the important reputation of the fern.

¶ Use

Says Gerard, 'Spleene-woort or milt-waste, being that kinde of Fern called Ceterach . . . beareth neither stalk nor seed . . . groweth upon old stone walls and rocks in darke and shadowie places throughout the West parts of England. Dioscorides teacheth that the leaves boiled in wine and drunke by the space of forty daies, take away infirmities of the spleen.' He takes exception to 'the gathering of Spleene-wort in the night, and other vain things, which are found here and there scattered in old books: from which most of the later Writers do not abstaine, who many times fill up their pages with lies and frivolous toyes.'

Culpeper recommends the herb for all 'melancholy diseases,' since 'Saturn owns it,' and he repeats the uses mentioned by other herbalists.

Ceterach has been used as bait for cod on parts of the Welsh coast because the rough scaly leaves give the plant a resemblance to certain kinds of worms and millipedes.

¶ Note

Comptonia peregrina (L.) Coulter (formerly *C. asplenifolia*), a plant known as sweet-fern and whose leaves bear a strong resemblance to ceterach, is not a fern at all but a shrub of the Bayberry Family. The leaves are fragrantly aromatic, tonic and astringent, and served as a tea substitute at the time of the Revolution. The genus is named for Henry Compton (1632–1713), Bishop of London and a patron of botany. Sweet-fern is a native of North America and ranges from Nova Scotia and Saskatchewan southward to North Carolina and Indiana. A leaf is illustrated below.

Agnus Scythicus
(*Dicksonia Barometz*).

Cibotium barometz (L.) J. Smith

Barometz, Scythian lamb, Tatarian lamb, Vegetable lamb

¶ *Description* Small tree fern to 15 feet. Leaves to 24 inches long, tripinnate, lanceolate or triangular. Leaflets narrow, acute, subdivided into long, acute, scythe-shaped lobes. Sori located on the lobes, close to the midvein of the leaflet. Rhizome or stem to 12 inches long, rough and irregular in shape, covered with yellowish-brown hairs. The natives of the regions where the fern grows have always compared the rhizome to the body of an animal.

¶ *Habitat and Distribution* Plains around the Volga, Asiatic Russia; Turkestan, China, India, Malay Peninsula. A plant of temperate to sub-tropical environment. Requires constant moisture for the rhizomes.

¶ *History* The Vegetable lamb is one of the items in the Seventeenth Tribe of John Parkinson's *Theatrum Botanicum*. The last division in the book, the Tribe consists of an ill-assorted miscellany which the author did not know how to dispose of elsewhere, including cacti, cardamums, cloves, civet, the unicorn's horn and the Indian dreamer (hemp). Of the Scythian lamb, he says it is 'reported by divers good authors, . . . is called Agnus Scythicus or *Planta animal*; it groweth among the Tartares about Samarcanda.' Tatars or Tartars were certain tribes inhabiting Manchuria and Mongolia. Samarkand in Turkestan is described by Marco Polo as 'a noble city adorned with beautiful gardens, and surrounded by a plain.'

The legend of a creature half plant, half animal, existing in the Orient, is of ancient origin, passed on for centuries by word of mouth. The tale was first published in a book about 1356, the *Travels* of the Englishman, Sir John Mandeville, whose account of his journeys into Turkestan and China is embellished with imaginary wonders and culled in part from Marco Polo and Saint Odoric, a Franciscan friar, born near Pordenone in northeastern Italy about 1286. Odoric traveled in Tartary as a missionary for about ten years.

It was generally accepted that the fabulous creature was a lamb, reported by some as growing inside a gourd, by others as having its legs rooted in the ground or its body supported and nourished by a rooted stalk. It was named barometz, from the Tatarian word.

In 1725 a German scientist first published the idea that the vegetable lamb was from a fern; soon after, Sir Hans Sloane, an English botanist and personal physician to George II, presented a specimen of the vegetable lamb to the Royal Society of London, whose president he was from 1727 to 1740. His extensive collection of specimens and books formed the beginning of the British Museum.

Some authorities have argued that cotton is the vegetable lamb, but there is nothing legendary about cotton, a familiar commodity from ancient times. The identity of barometz belongs to the fern, and is even now perpetuated by Chinese folk artists who carve 'lambs' from the fern stems and sell the figures to tourists.

The illustration on page 43, reproduced from a nineteenth-century engraving, is much less fanciful than earlier renderings. It shows the rhizome in an inverted position. The creature appears to have two tails and is missing a head. The caption indicates that barometz was formerly placed in *Dicksonia*.

¶ *Use* The rhizome, which has a sweetish-astringent flavor, is used in China as a tonic and as a remedy for rheumatic pains. The hairs are laid over wounds to staunch bleeding. As a decorative plant, the fern is very appropriate for large patios and lobbies.

The illustration on this page shows a subleaflet from the base of the leaf.

Dennstaedtia punctilobula

Dennstaedtia punctilobula (Michx.) Moore

Hay-scented fern, Boulder fern, Dicksonia, Gossamer fern, Mountain fern

Plant 2 to 3 feet high; fronds light green, ovate-lanceolate, acute, delicate. Leaflets lanceolate, acute, the lower surface covered with glandular hairs which are the source of the hay-like odor. Sori located at the indentations of the subleaflets. Indusium cup-shaped. Rhizomes slender, shallow, creeping and matted. The fronds die in autumn frost and turn whitish. ¶ *Description*

Grows equally well in shade or in the open and tolerates most soils, but prefers somewhat acid woodland humus. It is found commonly in hilly pastures among boulders and rock outcroppings, especially in New England. The fern is a North American species and ranges from Newfoundland and Minnesota southward to Georgia and Missouri. ¶ *Habitat and Distribution*

Has some of the weedy tendencies of bracken, spreading by extensive horizontal rhizomes, but the shallow root system is more easily controlled than the deep one of bracken. Does well in average woodland soil and does not require a great deal of moisture. Makes a decorative and easily maintained ground cover. ¶ *Culture*

Formerly called *Dicksonia punctilobula* for James Dickson (1738–1822), an English horticulturist, the hay-scented fern is now placed in the genus named after August Wilhelm Dennstaedt (b. 1800), a German botanist and author of several books on plants. ¶ *History*

Thoreau epitomizes the virtue of the fern which he knew by its old name: 'Is there any essence of Dicksonia, I wonder? When I wade through by narrow cow-paths, it is as if I had strayed into an ancient and decayed herb garden. Nature perfumes her garments with this essence now especially. She gives it to those who go a-berrying and on dark autumnal walks. The very scent of it, if you have a decayed frond in your chamber, will take you far up country in a twinkling. You would think you had gone after the cows there or were lost on the mountains.'

In herb potpourri, bouquets, sachets, closets, chests. ¶ *Use*

Dryopteris filix-mas

Dryopteris filix-mas (L.) Schott

Male fern, Male shield fern, Knotty brake, Shield root

Fr.: Fougère male   It.: Felce maschia
 Ger.: Johanniswurzel, Wurmfarn

A large leathery fern with fronds up to 5 feet long. Leaves broadly lanceolate, divided into numerous pairs of long, acute leaflets. Leaflets divided into numerous slightly ascending subleaflets with serrate margins. Stalk short, scaly. Sori large, with kidney- or shield-shaped indusium. Rootstock thick, erect and scaly. ¶ *Description*

Common throughout Europe and the British Isles. Rare in the United States; ranges from Maine and northern Vermont westward to Michigan. Also found in temperate Asia, parts of Africa, and in Mexico. It grows in limestone areas among rocks in damp shady woodlands. ¶ *Habitat and Distribution*

Male fern is commercially grown for medical use as an anthelmintic. Rhizomes not under three inches long are dug in autumn, trimmed of scales and roots, and split in half lengthwise. If marketed dry, the pieces are dessicated at a constant temperature of 70 degrees; some are locally sold fresh. ¶ *Culture*

Cultivated varieties with crested fronds are most desirable for landscaping and are available from nurseries. They require the same conditions as the wild plant: shade and moist, neutral woodland soil.

¶ *History* Known since ancient times as a reliable vermifuge, specifically for tapeworm, male fern is recommended by both Theophrastus and Dioscorides, who say that no part of *Pteris* (Greek for wing) is useful except the rootstock, which drives out the flat-worm. Although well known, the rhizome in powdered form mysteriously found its way into a secret formula which was sold to Louis XVI of France by Madame Nouffer, widow of a Swiss surgeon, for a sum equivalent to about $3500.

The 'seed' of male fern was eagerly sought for its property of making the bearer invisible. The seed being invisible, it bestowed invisibility. Fugitives or hunted animals disappearing among ferns vanished because fern seed fell on them. Seed of the thickest and tallest ferns had such power.

On Midsummer Eve, the eve of the Feast of St. John the Baptist (June 24), it was customary in England to kindle fires called St. John's fires on the hilltops in celebration of the summer solstice. The wild festivities, rituals and superstitious practices then observed gave rise to the expression 'midsummer madness.' On the Eve of St. John fern seed was gathered by moonlight in various complicated and elaborately planned procedures, one of which called for the use of twelve pewter plates, an opportunity for thieves or to impress the neighbors. St. John's wort was burned on Midsummer Eve; male fern was gathered and sold to be worn as a charm or to put into water that cows drank to protect them from spells. From rootstocks and dried fiddleheads sellers of relics and charms formed shapes called 'St. John's hands' which they sold to the gullible.

For the historical explanation of the fern's epithet, 'male,' see page 30.

¶ *Use* *D. filix-mas* is the only fern ever included in the United States Pharmacopoeia. Long known in Europe and Asia, the herb is currently sold in apothecary shops of modern China, India and other countries. The principle, extracted from the rootstock, is a thick dark green liquid called Oleoresin of Male Fern. It is poisonous if not taken in carefully regulated doses. Forms used are powder, fluid extract or extract by ether, the latter being the most effective, containing at least a fourth of crude filicin. In Gerard's time the bruised rhizome was 'drunke in Mede or honied water; and more effectually if it be given with two scruples or two third parts of a dram of Scamonie [Syrian bindweed], or of blacke Hellebor: they that will use it . . . must first eate Garlicke.' Gerard was not taking chances. Garlic, scammony, and hellebore are all strong anthelmintics.

Dryopteris fragrans

Dryopteris fragrans (L.) Schott

Fragrant fern, Fragrant cliff fern, Fragrant shield fern

Leaves 6 to 20, in a crown. Dried leaves of preceding seasons persist. Blades lanceolate, 3 to 12 inches long. Leaflets deeply pinnatifid, lanceolate. Leaflets vary in size and form depending on exposure and elevation. Two types of leaves are illustrated, the plate opposite showing a plant collected at an elevation of 4000 feet. Rootstock stout, erect, scaly. Sori large, nearly round, often covering entire under surface of leaves, giving them a brown color. ¶ *Description*

Northern New England to Wisconsin and northward. In the White Mountains, the Green Mountains, and in the Adirondacks. In crevices of dry, shaded rocks and cliffs, and on talus slopes. Requires a calcareous soil and a somewhat dry environment. Rare in Europe except in the extreme north, where it ranges from Greenland to Siberia. ¶ *Habitat and Distribution*

The fern has astringent properties and a fragrance which has been described as resinous or as resembling raspberry or primrose. The odor of the dried fronds persists for years. The dried herb is used in the same ways as the hay-scented fern, but the fragrant fern is not so easy to obtain because of its more remote habitat. ¶ *Use*

In Yakutsk, a province of Siberia on the Arctic Ocean, the fragrant fern is used for tea. The people, called Yakuts, are of mixed Turkish origin and inhabit the region along the Lena, a major river of Siberia.

Dryopteris spinulosa

Dryopteris spinulosa (O. F. Muell.) Watt

Spinulose wood fern, Prickly-toothed shield fern, Narrow buckler fern, Toothed wood fern, Fancy fern, Florist's fern

Fr.: Polystic spinuleux   Ger.: Dornfarn, Dorniger Schildfarn

Plant 1 to 3 feet high, growing in a tuft from the crown of a thick, horizontal rootstock. Stalk with light brown scales. Fronds lanceolate, delicately cut. Leaflets ascending, lowest pair obliquely triangular with pinnules longer on lower side, the longest next to the axis. Pinnules with serrate segments tipped with minute spines. Sori in rows on veins of pinnules. Indusium smooth, kidney-shaped. Leaves extremely variable, with numerous named varieties. ¶ *Description*

From Newfoundland to Alaska. In the Eastern United States southward to Virginia; rare south of Pennsylvania. Common in central Europe, mountains of southern Europe, the British Isles and north temperate western Asia. In wet shady woodlands, swamps, and near mountain springs in neutral or slightly acid rich humus. ¶ *Habitat and Distribution*

Adaptable to various types of rich soil, but requires constant moisture and damp mulch in midsummer. Effective as mass planting on shaded slopes. ¶ *Culture*

The rhizome is used in Europe, especially in Finland and Sweden, as a remedy for tapeworm in combination with male fern, since the spinulose wood fern has the same properties, but to a lesser degree. ¶ *Use*

The natives of northern Asia and of this continent bake and eat the fiddleheads which have a slightly sweetish taste.

The fronds are the 'Fancy Fern' of the florist trade. New England, where the fern is abundant, has been the major source. The quantities taken for cold storage tended to deplete the population. The fern is now cultivated.

Matteuccia struthiopteris (L.) Todaro

Ostrich fern

¶ *Description* Sterile leaves to 5 feet long, shaped like an ostrich plume, widest near the top. Leaflets long, narrow, acute, ascending. Spores on separate, shorter leaves, becoming dark brown at maturity. Fertile leaflets tightly wrapped around sporangia, forming thick rigid segments. After spores are shed, leaflets appear more feathery and persist dried through winter. In the fiddleheads of sterile leaves, the pinnae project outward from the axis, as illustrated above.

¶ *Habitat and Distribution* The ostrich fern is the only species of *Matteuccia* in North America. It ranges from Newfoundland to Alaska, and south to Virginia and Missouri. Rare in Europe, appearing only in the cool northern regions. Grows in partially sunny wet areas, such as banks of streams, swamps, and low wet woodlands.

¶ *History* Formerly called *Pteretis pensylvanica*. The present generic name is after Carlo Matteucci (1800—1868), an Italian physicist.

¶ *Culture* The graceful form, large size and rich color have made the ostrich fern a favorite for landscape planting in large masses. Needs highly acid soil and moist shady conditions, with frequent watering to maintain size. Spreads rapidly.

¶ *Use* The Indians of eastern Canada and New England showed the colonists how to boil and bake the rootstocks and cook the fiddleheads as greens. The sterile fiddleheads of ostrich fern are rated the best for culinary use. Packages of frozen fiddleheads are available in grocery stores in large urban centers. A good-sized planting in the garden provides fresh fiddleheads, which may also be put in the freezer.

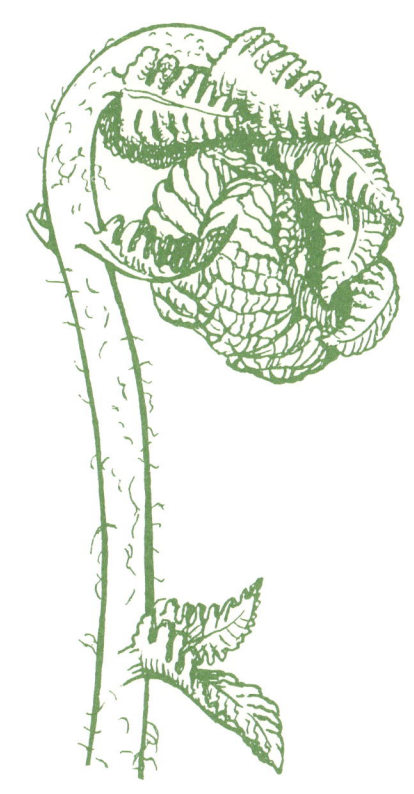

Osmunda claytoniana L.

Interrupted fern, Clayton's fern

¶ *Description* Leaves oblong, lanceolate, leathery, arching; may grow to 5 feet or more in wet acid soil. Leaflets lanceolate, ascending, deeply cut into overlapping lobes. Fertile leaves taller than sterile, interrupted near center of axis by several pairs of leaflets bearing sporangia. Fertile leaflets turn brown when spores are ripe. A fiddlehead is illustrated above. The tall greenish-white ones of *O. claytoniana* and *O. cinnamomea* are difficult to tell apart, and the mature plants look similar, an important difference being the separate fertile fronds of the cinnamon fern.

¶ *Habitat and Distribution* Interrupted fern is not particular as to soil or other conditions but is more likely to be found in somewhat dry borders of woodlands. Grows large in wet areas. From Newfoundland to eastern Manitoba, south to Georgia and Arkansas. Cinnamon fern prefers wet swampy conditions.

¶ *Use* The wiry matted roots of *Osmunda* species are used as a medium for holding the tubers of orchids in greenhouse culture.

The fiddleheads of these two species are edible, as the settlers learned from the Indians. Fiddleheads now rank among 'gourmet' foods and are prepared and served like asparagus but may be carcinogenic.

Osmunda regalis

Osmunda regalis L.

Royal fern, Water fern, Flowering fern, Bog onion, Buckhorn brake, Ditch fern, French Bracken, Heart of Osmund, Herb-Christopher, Hog brake, King fern, Locust fern, Osmund royal, Osmund-the-waterman, Snake brake

Fr.: Osmonde, Fougère fleurie, Fougère aquatique, Fougère royale
Ger.: Königsfarn, Osmunder   It.: Felce florida

¶ *Description*

Height of plant 2 to 6 feet. Leaves bipinnate, the spacing and shape of the leaflets giving the fronds a resemblance to the foliage of the locust tree. Six or more pairs of opposite, oblong, ascending leaflets. Subleaflets alternate, oblong with rounded or slightly pointed tips and oblique bases; terminal subleaflets sometimes larger and eared. Fertile branches at top of main axis bear masses of sporangia, the whole resembling an inflorescence and giving rise to the name 'flowering fern.'

Gerard gives this descriptive comment: 'When I first saw them afar off it caused me to wonder thereat, thinking that I had seene young Ashes growing upon a bog.' Ash trees and locusts also have compound leaves. The rhizome of royal fern is thick, matted with scales and roots, and may exceed six inches in height. Stipes stiff, reedy, brown.

In Culpeper's description the fern 'shooteth forth in spring-time divers rough hard stalks, half round: having divers branches of winged yellowish green leaves, set one against another and not nicked on the edges. From the top of these stalks grow forth a long bush of small green scaly aglets which are accounted the flowers and seeds.' (An aglet was an ornamental metal tip on a ribbon or lacing, similar to the tip on a modern shoe string.)

Common throughout northern Europe and the British Isles, ¶ *Habitat and* particularly in the west of England and Scotland, in Wales and *Distribution* parts of Ireland. In the western hemisphere, from Newfoundland and southern Canada southward to Florida and Tennessee. In swamps, bogs, low wet meadow land and along the margins of streams and ponds. In true bog conditions of high acidity and no drainage *Osmunda regalis* may grow to 6 feet.

Parkinson rightly observes that the royal fern was not known ¶ *History* to the ancients, because it is a plant of northwestern Europe. Osmunder was the name of the Saxon god equivalent to the Scandinavian Thor and to Vulcan, smith of the gods in classical mythology. In metallurgy of early medieval Europe, iron was made directly from bog iron ore found on pond bottoms and in swampy areas. Pieces of ore, as well as the rough masses of iron extracted in a primitive forge, were called osmunds. Bog iron is an impure ore called limonite, formed since prehistoric times by the deposit of iron in the water on remnants of animal organisms and decayed vegetation. The association of royal fern with the localities where bog iron is found and with fossil impressions discovered in the areas may account for the plant's name.

Another explanation of local legendary origin is the story of a waterman named Osmund who saved his wife and daughter during a Danish invasion by hiding them among the ferns of Loch Tyne. The ferns were later named after him. In Saxon the given name Osmund signified the domestic peace of the hearth, from the god Osmunder's association with fire, so the tale may be an allegory.

Early New England foundries depended greatly on bog iron. Its discovery near Saugus, Massachusetts, led John Winthrop, Jr. to establish works there about 1643. Other furnaces were started later, including one in partnership with Ethan Allen.

'Saturn owns the plant,' says Culpeper. 'The decoction to be drank ¶ *Use* or boiled into an ointment of oil, as a balsam or balm, and so it is singular good against bruises, and bones broken . . .' According to Gerard, 'Most of the Alchimists call it *Lunaria maior* . . . it can dissolve cluttered bloud . . . The root is great and thicke having in the middle some small whitenesse which hath been called the heart of *Osmund* the water-man.'

This white inner part of the rhizome is edible; its slightly pungent flavor gave it the name bog onion; it is said to have tonic properties from chemical assimilation of minerals from bog water.

Ophioglossum vulgatum

Ophioglossum vulgatum L.

Adder's tongue, Adder's spear, Adder's spit, Christ's spear, Serpent's tongue
Fr.: Langue de Serpent, Herbe aux Cent Miracles, Herbe sans Couture
It.: Erba Luccia

¶ *Description* Plant 3 to 12 inches high, succulent. The single blade simple oblong-oval, veins netted. Sporangia in two rows at top of long sporophyll stalk. Base of stipe sheathes bud of the next season. Gerard's words, 'One leaf and no more, fat or oleous in substance, from the bottom of which leaf springeth out a small and tender stalke, on the end of which doth grow a long small tongue.'

The generic name of the herb is from Greek, meaning serpent's tongue.

Some modern botanical authorities are of the opinion that *Ophioglossum* and *Botrychium* (see pages 33 and 37) should not be classified with the ferns, but should form a separate order.

¶ *Habitat and Distribution* Common throughout the United States, Europe and the British Isles. Also in temperate Asia, Africa and Australia. Grows in damp loamy or sandy turf, in grassy meadows and open woods in limestone regions. Prefers subacid soil; tolerates some dryness and sunlight.

¶ *Use* Principally used as a vulnerary in the form of an ointment called 'Green Oil of Charity,' made by combining 2 pounds of chopped leaves with one-half pint of oil and one and a half pound of suet, the whole boiled together until the herb material becomes crisp; the mixture is then strained. The leaves have sometimes been boiled with unsalted butter. In Gerard's directions, 'The leaves of Adder's tongue stamped in a stone mortar and boyled in Oile of Olive until the herbe be dry and parched, and then strained, will yeeld a most excellent greene oyle, or balsame . . . whose beautie is such that very many Artists have thought the same to be mixed with Verdigrease.' Verdigris, or acetate of copper, is the green substance formed on copper utensils in the presence of vegetable acids. It is used as a pigment in the arts, dyeing, for example.

In *New England's Rarities* Josselyn says, 'Adder's tongue comes not up till June; I have found it upon dry hilly grounds . . . in August, and did then make Oyntment of the Herb new gathered.'

The juice of the herb, pressed or distilled, is an old remedy for sore eyes, and for nosebleed and internal bleeding, such as from ulcers.

'Adder's tongue is under the Moon and Cancer [June],' declares Culpeper, 'and so cures by sympathy all diseases in parts of the body governed by the Moon & by antipathy those parts governed by Saturn.'

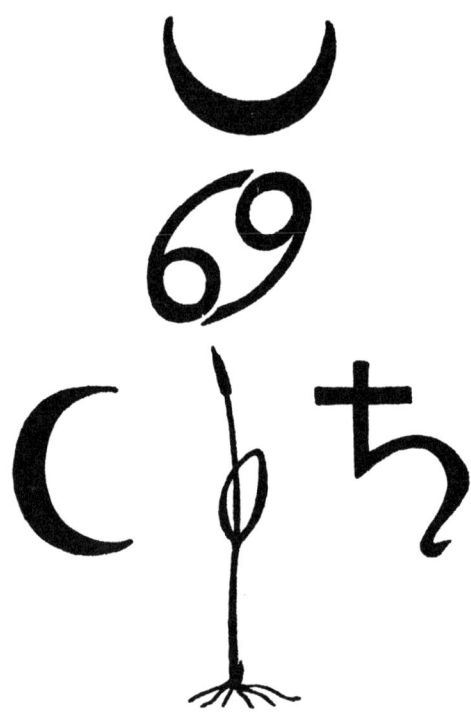

Phyllitis scolopendrium (L.) Newm.

Hart's-tongue fern, Burnt-weed, Buttonhole, Caterpillar fern, Finger fern, Christ's hair [Island of Guernsey], Fox tongue, Hind's tongue, Horse tongue, Lamb's tongue, Seaweed fern, Snake-leaves

Fr.: Langue de cerf, Herbe a la rate
Ger.: Hirschzunge   It.: Lingua cervina, Lingua da pozzi

¶ *Description*   Blades entire, linear or oblong-lanceolate, leathery, bases cordate or eared, margins wavy, maximum length about 3 feet, but usually about 12 inches or less. Stalk short, with brown scales. Sori linear, in pairs along veins. Rhizome compact, chaffy with brown scales. The plant grows in masses, spreading by new crowns in circular formation.

¶ *Habitat and Distribution*   Common throughout central and northern Europe and the British Isles. In the mild damp climate of Ireland and the Channel Islands the fern reaches maximum size. Grows also in Russia, Greece, Spain, Asia Minor, Azores, North Africa and Japan. Very rare in the Western Hemisphere, originally found only in a small area of north-central New York and two areas in New Brunswick, Canada. Since about 1930, efforts have been made to naturalize the fern in various localities in the United States. Prefers rocky limestone areas and neutral soil, damp shady crevices, old walls and ruins. Sunlight is tolerated but causes diminution in length of fronds.

¶ *Culture*   Imported plants may be obtained from dealers. Crested varieties were developed in England more than a century ago. Their requirements are the same as the wild plants.

¶ *History*   The herb appears in manuscripts of Apuleius (c. 1100 A.D.) as Hertestunge, Scolopendrio, Herba splenion, A splenis dolorem. First called hirtzunge by St. Hildegarde of Bingen, abbess of the Benedictine convent at Disibodenberg from 1136 until her death in 1179. She was noted for great learning, and her visions came to the notice of Bernard of Clairvaux when he was preaching the Second Crusade. St. Hildegarde's book, the

Physica (Natural Science), contains much on the practice of medicine, ancient remedies, contemporary folk lore, and her own knowledge.

Skolopendra is Greek for centipede, and the specific name refers to the rows of sori whose appearance suggested the name.

¶ *Use* Powdered dried root recommended by the Greek physicians, Galen and Dioscorides, for its astringent property as a remedy for dysentery.

In Goodyer's version of Dioscorides, 'the plant is also helpful for 4-footed beasts being poured in through ye mouth.' Galen, who learned surgery by volunteering to patch up gladiators of the Roman arenas, recommends the herb as a vulnerary. 'Jupiter,' proclaims Culpeper, 'has dominion over this herb, therefore it is a singular remedy for the liver . . . you shall do well to keep it in a syrup all the year. Hart's Tongue is much commended against the hardness and stoppings of the spleen. The distilled water thereof is also very good against the passions of the heart, and to stay the hiccough.'

The mucilaginous and astringent properties of the herb provide a quick remedy for burns, scalds, eruptions and wounds. The raw bruised leaves are applied either alone or mixed with grease. The fern has been used officinally in combination with *Hydrastis canadensis* (golden seal, orange root) for various internal disorders.

Polypodium vulgare

Polypodium vulgare L.

Common Polypody, Moss fern, Oak fern, Rock-cap fern, Snake fern, Stone fern, Wall fern, Brake root, Golden Locks, Polypody-of-the-oak

Fr.: Polypode vulgare, Polypode de chène, Réglisse sauvage, Tripe de roche
Ger.: Engelsüss, Engelwurz, Gemeiner Tupfelfarn
It.: Erba radioli, Felce dolce, Polipodio

¶ *Description* Fronds oblong-lanceolate to triangular, leathery, light green, to 12 inches in length. Blade divided into 10 to 20 pairs of leaflets with blunt to sub-acute tips. Sori large, round, located at ends of free veins, mostly on upper leaflets; indusium lacking. Rhizome horizontal, scaly.

¶ *Habitat and Distribution* Common in Europe and the British Isles and in most parts of the United States. In shade or partial shade, on rocks and boulders or in crevices of cliffs, in neutral or moderately acid woodland soil. Sometimes on decayed logs, stumps or tree trunks. Thrives in shady damp areas, but tolerates some dryness.

¶ *Culture* Requires good drainage in light woodland humus among limestone rocks. Best transplanted with large quantity of original soil.

¶ *History* The Greek name *Polypodion* means 'many feet' and refers to the traces of the many stalks on the rhizome.

Because the ancients venerated the oak as sacred, they considered the polypody which grew on an oak to have special powers superior to those of the polypody that grew on the earth or rocks. The Druids of ancient Britain held their religious ceremonies in oak groves and venerated any plant which grew on the oak, including mistletoe and polypody. They helped to perpetuate a belief in two species of polypody. We now know that the two are the same species, but the early physicians argued about the distinction for centuries. Culpeper holds forth, 'Polypody of the oak groweth upon old rotten stumps or trunks of trees, as oak, beech, hazel . . . That which groweth upon oak is accounted the best; but the quantity thereof is scarce sufficient for the common use. And why, I pray, must Polypodium of the Oak only be used, gentle college of physicians? It is only because it is dearest. Will you never leave your covetousness till your lives leave you? The truth is, that which grows upon the earth is best ('tis an herb of Saturn, and he seldom climbs trees) to purge melancholy; if the humour be otherwise, chuse your Polypodium accordingly.'

¶ *Use* Principal uses tonic, pectoral, purgative. An important remedy for children's diseases, often fatal, including whooping cough, asthma and tuberculosis. A mucilaginous expectorant was made by boiling the fronds in sugared water or by mixing the dried powdered rhizome with honey. Bock in *De Stirpium Historia* (1552) called the herb and the remedy 'angel-sweet' and also 'wild licorice,' from the sweetish taste of the rhizome. Bock also recommends the powdered rootstock in a chicken stew with ginger and anise added, as a remedy for ulcers and the black bile. This is a variation of the oft-repeated recipe of Dioscorides, which includes also fish, beets or mallows. Culpeper recommends the distilled water of rootstock and leaves 'for the quartan ague [malaria], also against melancholy, or fearful and troublesome sleeps or dreams . . .'

'For a certain kinde of *Arthritis*, or ache in the joints,' says Gerard, 'it is very much commended by the Brabanders and other inhabitants about the river Rhene (Rhine) and the Maze (Meuse).' Brabant, an ancient duchy, in modern times became part of the Netherlands and Belgium.

In a quite different usage, Izaak Walton advises in *The Complete Angler* (Part I, Chapter VII, concerning trout and salmon): 'Take the stinking oil drawn out of polypody of the oak by a retort, mixed with turpentine and hive-honey, and anoint your bait therewith.'

Like bracken (see the next section), common polypody contains a large amount of carbonate of potash and was formerly used in the manufacture of glass.

Pteridium aquilinum

Pteridium aquilinum (L.) Kuhn

Bracken, Brake, Eagle fern, Erne fern, Pasture Brake, Umbrella fern

Fr.: Fougère commune, Fougère impériale, Aigle impériale
Ger.: Adlerfarn, Minutenkraut   It. Felce aquilina, Felce da ricotte

¶ *Description* Plant usually 2 to 4 feet high, but in favorable conditions may reach 10 feet or more. Leaves triangular to ovate, dull green, leathery, divided into three approximately equal parts, the two lower opposite and divided into leaflets and subleaflets, the upper divided into leaflets. Sori continuous along margins of leaflets. Rootstock often many feet long, deep, horizontally creeping. When the lower portion of the stalk is sliced in cross section, the tissues show a pattern, as illustrated in the plate opposite. Various interpretations have been given to the pattern. Hieronymus Bock mentions the appearance of a double-headed eagle in the stem. Others are described below under *History*.

¶ *Habitat and Distribution* Common in Europe, the British Isles and the United States, but not found in wet or calcareous regions. Does well in poor sandy soil in partial shade of open woods, and grows rankly in full sun in old pastures, abandoned fields and other waste land. Associated with heather and gorse on the sandy moors of England and Scotland.

¶ *Culture* Not desirable for cultivation because of its weedy character. The fronds may be used as a winter mulch, but their low nitrogen content gives them little value as fertilizer. Bracken ashes, however, are a source of potassium.

¶ History

Bracke in medieval German meant 'breaking of ground after harvest.' In Old Dutch *brachfeld* meant fallow land. In Old English brake meant land broken but not sown, rough ground, brushland or thicket overgrown with shrubs or brambles. 'This green plot shall be our stage, this hawthorn brake our tiring house.' (*Midsummer Night's Dream*, Act III, scene 1.)

Among the interpretations of the figure in the cross-section of the stem, one was called 'King Charles in the oak,' a reference to the story that King Charles II of Scotland escaped his enemies by hiding in an oak tree. Another Scotch legend gave the figure the name 'Devil's Foot.' In Ireland the brake is sometimes called Fern of God, because the cut stalk shows the letters G, O, D. Others have seen the letter C for Christ, or X, the Greek initial of Christos. These symbols gave the brake repute as a charm against evil.

It was said that on St. John's Eve the fern produced a blue flower which ripened at midnight into a shining seed. Caught in a white napkin, the seed had power to render the possessor invisible. (See also *Dryopteris filix-mas*.) In Germany the fern was believed to bloom for only one minute, and in Russia the seed was believed to confer second sight. In Swabia it was said that fern seed brought by the devil at midnight enabled one man to do the work of thirty, a tale that may account for the name 'back-ache fern.'

¶ Use

The brake has anthelmintic properties similar to male fern but is not officinal. The high tannic acid content has sometimes proved to be toxic to grazing cattle.

In Europe in times of poor grain harvest the rhizomes of bracken were formerly ground as an extender for flour, particularly in eighteenth-century France, when the peasantry, driven to economic ruin by the extravagance of the nobility, resorted to the consumption of any vegetation that provided

even minimal nourishment. The rhizomes and young fronds were then and later boiled for use as pig feed, and in dried condition were added to cattle fodder. In the north of Europe, one part of bracken rhizome to two parts of malt was sometimes used in brewing beer. Gerard says that 'the root cast into an hogshead of wine keepeth it from souring.'

The fiddleheads are edible but not equal to those of the ostrich fern or species of *Osmunda*, and they need thorough cooking for digestibility. All parts of bracken have been eaten by the American Indians, who introduced the fern's culinary uses to the colonists.

Since ancient times bracken has been preferred as barn litter for livestock, especially sheep and goats, as recommended by the most famed of Roman agricultural writers, Vergil and Columella. The latter sets July and August as the best time for cutting bracken to check the spread of the fern. To Roman farmers bracken was a weed, for it invaded land that was suitable for vineyards, and the primitive wooden ox-drawn plows were inefficient in tearing up the strong creeping rhizomes. Pliny enlarges on how to kill bracken, and his translator, Philemon Holland, says, 'It will die at the root in two years if you will not suffer it to branch and grow above ground: knap off the head of the first spring with a wand or walking staff.' No great system for several acres, but a sound and effective principle.

Besides barn litter, bracken had another little known but major use as a ground covering and bedding for the soldiers in the Roman camps.

Glass is made by fusing silica with an alkali, either soda or potash. The alkali causes the silica to melt at a lower temperature, rendering the material workable for a longer time. The alkali in Egyptian, Greek and Roman glass was hydrate of sodium from ash of saltwort (*Salsola*) or glasswort (*Salicornia*), both seashore plants. The product became known as soda glass. Later in Roman times fern ashes came into use, as described by Eraclius in the twelfth-century work *De coloribus et artibus Romanorum*. In the rapid development of the arts and architecture in medieval Europe, the potentials of local material were quickly grasped. Ash of the brake was found to be rich in alkali, and its use became general in glass manufacture. The alkali is hydrate of potassium, and the glass is called potash glass. From Chaucer's *Canterbury Tales* (1388) comes the comment, 'But natheless, some said that it was wonder to make of fern ashes glass, and yet is glass not like ashes of fern.' In *De la Pirotechnia* (Venice, 1540), Vannoccio Biringuccio describes the use of both saltwort and fern in glass making. Parkinson mentions that ashes of fern were used in 'a kinde of thicke or dark coloured glasse in sundry places in France, out of which they drink wine.' Such glass was called 'verre fougère,' bracken glass.

In Europe prior to the Norman Conquest bracken ashes, alone or mixed with fats, were used for washing. Until the mid-nineteenth century, balls of fern ash were in demand for laundering clothes. For bleaching textiles, lye was made by mixing bracken ash with unslaked lime.

As a quick-burning high-heat fuel, bracken had specialized uses in lime-kilns, bakeries, breweries and brick-kilns.

Into the nineteenth century bracken was harvested in England for thatch, the entire plant being used, including the rhizome. Throughout the temperate world the fronds have been used as packing for fresh fruits, vegetables and fish. American Indians of the west coast split the rhizomes into bands for use in basket work.

Bracken presented problems to sheep raisers in Britain by taking over grassland needed for pasture. On the other hand, as cover for game, bracken was preserved by the nobility and the crown, and cutting and burning were strictly regulated. When burned for potash, the ashes were collected as fast as possible and stored in a dry place, because potash from bracken is very soluble, and the ash should not be exposed to rain. From the fear of rainfall at the burning of bracken came a belief in the constant imminence of the hazard. Consequently, when rain was much needed the bracken was fired to bring rain!

II
The Horsetails

The Horsetails

The Horsetails, also called Scouring Rushes, are the only living descendants of the tree-size *Calamites* which flourished about three hundred million years ago in the Carboniferous Age. The giant ferns, horsetails, and club mosses were the greatest coal-forming plants of prehistoric times. Cannel coal is composed principally of the petrified and carbonized spores and spore cases of these plants. The absence of woody material gives the coal a uniform texture, and it burns with a bright flame, 'like a candle,' cannel being a British dialectical form of candle. Cannel is also called soft coal, as opposed to hard or anthracite coal.

The old colloquial Latin name meaning horse bristle was given the genus by Linnaeus in 1753. The plants constitute the order *Equisetales*, the most primitive relatives of the ferns. They have stems, branches, and sheathing scales (leaves) at the nodes. All of the parts except the fertile stems manufacture chlorophyll. The stems are reed-like, with joints (nodes), and are hollow, with a large central tube or canal surrounded by smaller tubes. As seen in cross section these vary in the different species and are important factors in identification. Some species are unbranched. When present, the branches rise from the nodes in whorls. The plants have rhizomes and roots, the former sometimes many feet long. In certain species the epidermis of the stem contains a large amount of silica.

The fertile stem produces at the tip a strobilus or cone, bearing close-fitting scales which hold and cover the spore cases. At maturity the scales separate, and the spores are released. The spores are more or less globular and contain chlorophyll. They produce male and female prothalli, whereas ferns have only one prothallus. From the fertilized egg cell a new plant develops.

Horsetails grow principally in the temperate zones and inhabit swamps and wet sandy places on margins of ponds, lakes and streams.

Silica, sometimes called silex, is an extremely hard mineral substance that occurs in various forms, the commonest being quartz which constitutes seashore sand. The silica content of some species of horsetail may be nearly a fifth of the fresh plant. The particles of silica are usually in rows along the grooves of the stems and branches, but the arrangement varies in different species.

Because of the silica content, horsetails were formerly in great demand in the arts and crafts and household occupations as abrasives, hence the name scouring rush.

Equisetum arvense

Equisetum arvense L.

Field horsetail, Meadow horsetail

Fr.: Prele des champs, Queue de rat   Ger.: Acker-Schachtelhalm
It.: Coda di cavallo, brusca

¶ *Description*   Sterile stems to 18 inches high, light green, furrowed, surface rough from the presence of silica. Branches simple, numerous, and create a bushy appearance. Plants extremely variable in form.

The fertile stems which come up in spring before the sterile appear are about six inches high, succulent, buff colored. They sometimes bear very short branches. The strobilus has a blunt tip and is about an inch long.

¶ *Habitat and Distribution*   From northeastern Canada southward through temperate regions of the United States. Common in the British Isles and western Europe. In poor sandy damp soil, in cinder fill along railroad tracks, and in waste ground along roadside ditches. Presence of the plant is considered by some to indicate the location of subterranean water.

¶ *History*   Dioscorides called the horsetails *Hippuris* (from Greek *hippos*, horse). Apuleius mentions the herb as *Hyppirum*. The herbalists had trouble distinguishing the different species, and Mattioli simply refers to the four major medicinal ones as 'Equisetum I, II, III, IIII.'

¶ *Use*   Dioscorides recommends juice of horsetail combined with acetum (vinegar) as a vulnerary. Culpeper advises the addition of juice of moonwort leaves to distilled water of horsetail. A fluid extract of the fresh sterile plant has been officinal in Britain under the name *Cauda equina*, as a remedy for dropsy, gravel and ulcers. The dried powdered herb and the ash have been used as astringents for a stomachic.

Equisetum fluviatile L.

Great horsetail, River horsetail, Swamp horsetail, Paddock-pipes
Fr.: Prele élevée   Ger.: Riesen Schachtelhalm

¶ *Description*

Fertile stems reach a maximum height of 10 inches, are unbranched, have clasping sheathes with sharp-pointed teeth. Strobilus about 2 inches long, blunt-tipped. Sterile stems to 4 feet or more in height, smooth, pale, with approximately twenty inconspicuous ribs. Branches numerous, simple, starting at upper half of stem.

Strobilus and cross section of stem are illustrated above.

¶ *Habitat and Distribution*

Nova Scotia to Alaska, south to Virginia and Nebraska. Common in Europe and the British Isles. In slow or still water with muddy clay loam bottom. Swamps, ponds, river margins and wet meadows.

¶ *Use*

Although not especially palatable, the young shoots have been cooked like asparagus or fried in flour and butter or oil. They were eaten as long ago as Roman times by the peasantry of Europe. According to Linnaeus the great horsetail is eaten by reindeer, and in his time the plant was harvested in Sweden for cattle fodder. Medicinally it has been used dry or fresh for its astringent and diuretic properties. In England horsetail has long been a rural remedy for nosebleed.

Equisetum hyemale L.

Scouring rush, Dutch rush, Bottle brush, Horse pipe, Pewterwort, Rough horsetail, Shave-grass

Fr.: Prele d'hiver   Ger.: Winter-Schachtelhalm, Zinnkraut
It.: Asprella, Pincheri de' legnaiuoli

¶ *Description*
Fertile and sterile stems alike, unbranched, 3 to 4 feet high, rough, with about 30 ribs with distinct rows of silica. Sheaths appressed to stem, soon fall off. Strobilus has pointed tip. Dioscorides gives a description, and his translator words it, 'One stalk, tender, like a reed, having continued joints, lying one upon another, as it were of a trumpet, and round about ye joints, small leaves like those of ye pine.'

¶ *Habitat and Distribution*
Throughout North America, Europe and temperate Asia. Along shaded streams and lake margins; on wet, rocky or sandy banks and railroad grading.

¶ *History*
The specific name *hiemale* is Latin, meaning 'of winter,' and it was given to this species because the plant is evergreen. Linnaeus spelled it *hyemale* when he named the horsetails in 1753.

The name Dutch rush came from the large importation of the plant from Holland, where it grew profusely. It was sold by the bundle in enormous quantities in many countries.

¶ *Use* An old remedy as an astringent and vulnerary, but the high silica content created a major use for the scouring rush as an abrasive. The plant was used domestically to polish pewter, to scour kitchen utensils and milk pails, and to scrub floors; it was used by arrow makers to rub the wooden shafts, by comb makers to polish their bone and ivory products, and by cabinetmakers and metal workers of all kinds. One of the earliest artisans to mention the scouring rush was twelfth-century Roger of Helmarshausen, a monk who wrote his treatise *On Divers Arts* under the name Theophilus.

He says, 'Horse-saddles and eight-man carrying chairs, that is, curtained seats, and footstools and other objects that are carved and cannot be covered with leather or cloth, should be rubbed with shave-grass as soon as they have been scraped with an iron tool; then covered with two coats of gesso and when dry, smoothed again with shave-grass. . . . grows up like a rush and is knobby. You should gather this in the summer and dry it in the sun.' [Translation from the Latin by Hawthorne and Smith, Chicago, 1963.]

Scouring rush has been similarly used by the American Indians, who also consider the smoke of the burning plant to have a disinfectant property.

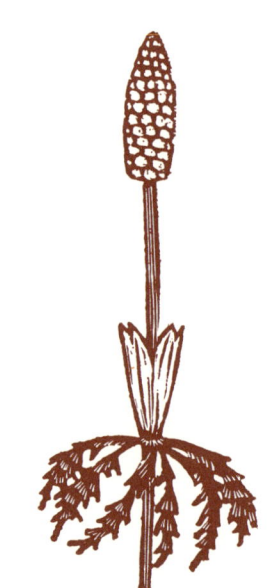

Equisetum sylvaticum L.

Wood horsetail, Branched wood horsetail

Fr.: Prele des bois   Ger.: Wald Schachtelhalm   It.: Rasperella

¶ *Description* Distinctive for bright green color and decumbent whorls of angular branches. Sterile plant resembles small evergreen tree.

¶ *Habitat and Distribution* Common in eastern United States, British Isles, northern and central Europe. Wet meadows and woodlands, swamps.

¶ *Use* Dioscorides recommends the herb finely crushed as a vulnerary, and to be taken in wine for nosebleed and dysentery.

III
The Club Mosses

The Club Mosses

The ancestors of the club mosses were big tree-like plants that flourished in the Paleozoic Era about three hundred million years ago. With the ancient ferns and horsetails, the club mosses contributed largely to formation of coal beds. The plants do not belong to the mosses, which constitute a different order.

The modern club mosses are small, mostly creeping, evergreen perennials. The stems and branches are covered with uniformly sized, overlapping, sharply pointed scales, arranged in opposite rows or in spirals. The plants reproduce by spores, but the most effective propagation is from the rhizome, which is of indeterminate extension.

The spores are borne in a terminal strobilus and ripen in late summer. The mature spore, when it falls on damp ground, develops a prothallus that bears the male and female elements whose union produces a new plant. The life cycle of *Lycopodium* is very long, involving several years instead of one or two seasons.

The name *Lycopodium* is Latin for wolf's foot, from an ancient fancied resemblance in the plant's manner of branching and creeping on the ground.

Lycopodium clavatum L.

Common club moss, Running club moss, Ground pine, Running pine, Forks-and-knives, Foxtail club moss, Muscus terrestris repens, Robin Hood's hatband, Staghorn club moss, Vegetable sulphur, Wolf's claw

Fr.: Lycopode en massue   It.: Erba strega, Stregonia
Ger.: Bärenmoos, Keuliger Bärlapp, Schlangenmoos, Wolfsranke

¶ *Description*
Stems creeping, often many feet long, rooted, sending up numerous irregularly forking branches. Scales small, arranged spirally on axis, evergreen, tapering to a fine bristle. Strobilus about 3 inches long, yellowish, on slender stem; the spore-bearing scales sharply tipped. In the Merck Index the spores are described as triangular pyramids with a convex base and rounded angles.

¶ *Habitat and Distribution*
From northeastern Canada west to Minnesota, south to Virginia, Kentucky and Iowa. Common in the British Isles and Europe. In open thickets and pine woods, on rocky hillsides, in light acid or subacid soil.

The specific name, meaning club-shaped, refers to the strobilus. ¶ *History*
Bock and other old writers called the herb *muscus terrestris*, earth moss, from its habit of growth.

In parts of Europe the plant was believed to create discord and argument when brought into a group of people. The spores have the curious property of being highly inflammable, and ignite explosively, which gave to the spore powder the name witch's flour.

Originally the whole plant was used dried as a diuretic for dropsy ¶ *Use*
and kidney ailments, as an aperient, a remedy for scurvy, a vulnerary and a cure for eczema. Bock recommends a decoction in wine for treatment of gout. The spores, which constitute a very fine soft powder, began to be used separately in the seventeenth century, also as a baby powder. They are officinal in the pharmacopoeias of Austria, Germany and Switzerland, for their oil and alkaloid content. The two latter countries and Russia are the principal sources. Called vegetable sulphur in the Merck Index, the powder is listed as used here principally for coating pills, since it repels water. It is also used in fireworks; in flash photography; as parting powder in foundry work because the spores are not only adsorbent, but are so fine that a mass of them flows almost like a liquid. Spores are shown below at magnified scale.

A remedy for steaming stiff joints has been made by the Indians from a decoction of *Lycopodium* combined with white spruce and ironwood.

Lycopodium complanatum L.

American club moss, American ground pine, Christmas green, Ground cedar, Flat-branch club moss, Running cedar

Fr.: Lycopode aplati   Ger.: Flachgedrückter Bärlapp, Jägergrün

Long, creeping, underground rootstock, from which rise ¶ *Description* numerous irregularly-branching stems. The flattened branches are covered with tiny, triangular, sharply pointed scales. Strobilus yellow, up to 3 inches long, on slender stems, scales sharply pointed. Plant evergreen, has aromatic odor and resinous taste.

Rocky hillside pastures, sandy edges of roadsides, borders of ¶ *Habitat and* pine woods, damp shaded woodland. Northeastern Canada, *Distribution* through New England, south to northern Pennsylvania; west to Minnesota and northern Iowa. Northern and central Europe and the Apennines.

Complanatum means flattened, descriptive of the branches. Gerard objected to ladies making coronets for themselves out of the plant.

¶ *History*

The whole plant, dried and powdered for infusion, served as a remedy for jaundice and splenetic ailments.

¶ *Use*

Appendix

Asparagus plumosus Baker

Asparagus fern. A woody climbing vine. Branches have twigs arranged in a horizontal plane, creating the semblance of a compound pinnate frond, triangular in outline, bright green. The species has many varieties, familiar additions to florists' bouquets. The plant is not a fern, but a member of the Lily Family and closely related to the culinary asparagus.

Myrrhis odorata Scop.

Sweet cicely, Sweet fern, Sweet chervil, Myrrh

   Not a fern, but a member of the Parsley Family. Perennial with a stout edible root, fern-like leaves, umbels of white flowers, and an aromatic anise-like flavor. Used as a pot-herb and as flavoring in soups, stews and salads. The old herbalists prescribed it as a stomachic and expectorant.

Pellaea mucronata D.C. Eaton

Bird's-foot fern, Cliff brake, Rock fern, Tea fern

   Fronds large, broadly deltoid-lanceolate, bi- to tri-pinnate, leaflets ascending. Herb evergreen, aromatic.

   Native to California. Dry rocky places; in the Sierra Nevada Mountains to elevations of 6000 feet. Used by prospectors and explorers as a tea.

   The generic name is from Greek *pellos*, meaning dusky, and refers to the dark brown stems of the herb. The name bird's-foot fern is a reference to the shape of the leaf in several species.

Library of Congress Cataloging in Publication Data

Abbe, Elfriede Martha, 1919–
  The fern herbal.

  "Reprinted, corrected edition"—T.p. verso.
  Reprint. Originally published: Manchester Center, Vt. : Press of E. Abbe, c1981.
  1. Medicinal plants. 2. Ferns. 3. Ferns—Therapeutic use.
4. Pteridophyta. 5. Pteridophyta—Therapeutic use. 6. Herbals.
I. Title.
QK99.A1A2 1985   587'.04634   84-45439
ISBN 0-8014-1718-X (alk. paper)